Ups and Downs

Britannica

ENCYCLOPÆDIA BRITANNICA EDUCATIONAL CORPORATION

Mathematics in Context is a comprehensive curriculum for the middle grades. It was developed in collaboration with the Wisconsin Center for Education Research, School of Education, University of Wisconsin–Madison and the Freudenthal Institute at the University of Utrecht, The Netherlands, with the support of National Science Foundation Grant No. 9054928.

National Science Foundation

Opinions expressed are those of the authors
and not necessarily those of the Foundation

ISBN 0-7826-1530-9
1 2 3 4 5 WK 02 01 00 99 98

The *Mathematics in Context* Development Team

Mathematics in Context is a comprehensive curriculum for the middle grades. The National Science Foundation funded the National Center for Research in Mathematical Sciences Education at the University of Wisconsin–Madison to develop and field-test the materials from 1991 through 1996. The Freudenthal Institute at the University of Utrecht in The Netherlands, as a subcontractor, collaborated with the University of Wisconsin–Madison on the development of the curriculum.

The initial version of *Ups and Downs* was developed by Mieke Abels and Jan Auke de Jong. It was adapted for use in American schools by Margaret R. Meyer, Julia A. Shew, Gail Burrill, and Aaron Simon.

National Center for Research in Mathematical Sciences Education Staff

Thomas A. Romberg
Director

Joan Daniels Pedro
Assistant to the Director

Gail Burrill
Coordinator
Field Test Materials

Margaret R. Meyer
Coordinator
Pilot Test Materials

Mary Ann Fix
Editorial Coordinator

Sherian Foster
Editorial Coordinator

James A. Middleton
Pilot Test Coordinator

Margaret A. Pligge
First Edition Coordinator

Project Staff

Jonathan Brendefur
Laura J. Brinker
James Browne
Jack Burrill
Rose Byrd
Peter Christiansen
Barbara Clarke
Doug Clarke
Beth R. Cole

Fae Dremock
Jasmina Milinkovic
Kay Schultz
Mary C. Shafer
Julia A. Shew
Aaron N. Simon
Marvin Smith
Stephanie Z. Smith
Mary S. Spence
Kathleen A. Steele

Freudenthal Institute Staff

Jan de Lange
Director

Els Feijs
Coordinator

Martin van Reeuwijk
Coordinator

Project Staff

Mieke Abels
Nina Boswinkel
Frans van Galen
Koeno Gravemeijer
Marja van den Heuvel-Panhuizen
Jan Auke de Jong
Vincent Jonker
Ronald Keijzer

Martin Kindt
Jansie Niehaus
Nanda Querelle
Anton Roodhardt
Leen Streefland
Adri Treffers
Monica Wijers
Astrid de Wild

Acknowledgments

Several school districts used and evaluated one or more versions of the materials: Ames Community School District, Ames, Iowa; Parkway School District, Chesterfield, Missouri; Stoughton Area School District, Stoughton, Wisconsin; Madison Metropolitan School District, Madison, Wisconsin; Milwaukee Public Schools, Milwaukee, Wisconsin; and Dodgeville School District, Dodgeville, Wisconsin. Two sites were involved in staff developments as well as formative evaluation of materials: Culver City, California, and Memphis, Tennessee. Two sites were developed through partnership with Encyclopædia Britannica Educational Corporation: Miami, Florida, and Puerto Rico. University Partnerships were developed with mathematics educators who worked with preservice teachers to familiarize them with the curriculum and to obtain their advice on the curriculum materials. The materials were also used at several other schools throughout the United States.

We at Encyclopædia Britannica Educational Corporation extend our thanks to all who had a part in making this program a success. Some of the participants instrumental in the program's development are as follows:

Allapattah Middle School
Miami, Florida
Nemtalla (Nikolai) Barakat

Ames Middle School
Ames, Iowa
Kathleen Coe
Judd Freeman
Gary W. Schnieder
Ronald H. Stromen
Lyn Terrill

Bellerive Elementary
Creve Coeur, Missouri
Judy Hetterscheidt
Donna Lohman
Gary Alan Nunn
Jakke Tchang

Brookline Public Schools
Brookline, Massachusetts
Rhonda K. Weinstein
Deborah Winkler

Cass Middle School
Milwaukee, Wisconsin
Tami Molenda
Kyle F. Witty

Central Middle School
Waukesha, Wisconsin
Nancy Reese

Craigmont Middle School
Memphis, Tennessee
Sharon G. Ritz
Mardest K. VanHooks

Crestwood Elementary
Madison, Wisconsin
Diane Hein
John Kalson

Culver City Middle School
Culver City, California
Marilyn Culbertson
Joel Evans
Joy Ellen Kitzmiller
Patricia R. O'Connor
Myrna Ann Perks, Ph.D.
David H. Sanchez
John Tobias
Kelley Wilcox

Cutler Ridge Middle School
Miami, Florida
Lorraine A. Valladares

Dodgeville Middle School
Dodgeville, Wisconsin
Jacqueline A. Kamps
Carol Wolf

Edwards Elementary
Ames, Iowa
Diana Schmidt

Fox Prairie Elementary
Stoughton, Wisconsin
Tony Hjelle

Grahamwood Elementary
Memphis, Tennessee
M. Lynn McGoff
Alberta Sullivan

Henry M. Flagler Elementary
Miami, Florida
Frances R. Harmon

Horning Middle School
Waukesha, Wisconsin
Connie J. Marose
Thomas F. Clark

Huegel Elementary
Madison, Wisconsin
Nancy Brill
Teri Hedges
Carol Murphy

Hutchison Middle School
Memphis, Tennessee
Maria M. Burke
Vicki Fisher
Nancy D. Robinson

Idlewild Elementary
Memphis, Tennessee
Linda Eller

Jefferson Elementary
Santa Ana, California
Lydia Romero-Cruz

Jefferson Middle School
Madison, Wisconsin
Jane A. Beebe
Catherine Buege
Linda Grimmer
John Grueneberg
Nancy Howard
Annette Porter
Stephen H. Sprague
Dan Takkunen
Michael J. Vena

Jesus Sanabria Cruz School
Yabucoa, Puerto Rico
Andreíta Santiago Serrano

John Muir Elementary School
Madison, Wisconsin
Julie D'Onofrio
Jane M. Allen-Jauch
Kent Wells

Kegonsa Elementary
Stoughton, Wisconsin
Mary Buchholz
Louisa Havlik
Joan Olsen
Dominic Weisse

Linwood Howe Elementary
Culver City, California
Sandra Checel
Ellen Thireos

Mitchell Elementary
Ames, Iowa
Henry Gray
Matt Ludwig

New School of Northern Virginia
Fairfax, Virginia
Denise Jones

Northwood Elementary
Ames, Iowa
Eleanor M. Thomas

Orchard Ridge Elementary
Madison, Wisconsin
Mary Paquette
Carrie Valentine

Parkway West Middle School
Chesterfield, Missouri
Elissa Aiken
Ann Brenner
Gail R. Smith

Ridgeway Elementary
Ridgeway, Wisconsin
Lois Powell
Florence M. Wasley

Roosevelt Elementary
Ames, Iowa
Linda A. Carver

Roosevelt Middle
Milwaukee, Wisconsin
Sandra Simmons

Ross Elementary
Creve Coeur, Missouri
Annette Isselhard
Sheldon B. Korklan
Victoria Linn
Kathy Stamer

St. Joseph's School
Dodgeville, Wisconsin
Rita Van Dyck
Sharon Wimer

St. Maarten Academy
St. Peters, St. Maarten, NA
Shareed Hussain

Sarah Scott Middle School
Milwaukee, Wisconsin
Kevin Haddon

Sawyer Elementary
Ames, Iowa
Karen Bush Hoiberg

Sennett Middle School
Madison, Wisconsin
Brenda Abitz
Lois Bell
Shawn M. Jacobs

Sholes Middle School
Milwaukee, Wisconsin
Chris Gardner
Ken Haddon

Stephens Elementary
Madison, Wisconsin
Katherine Hogan
Shirley M. Steinbach
Kathleen H. Vegter

Stoughton Middle School
Stoughton, Wisconsin
Sally Bertelson
Polly Goepfert
Jacqueline M. Harris
Penny Vodak

Toki Middle School
Madison, Wisconsin
Gail J. Anderson
Vicky Grice
Mary M. Ihlenfeldt
Steve Jernegan
Jim Leidel
Theresa Loehr
Maryann Stephenson
Barbara Takkunen
Carol Welsch

Trowbridge Elementary
Milwaukee, Wisconsin
Jacqueline A. Nowak

W. R. Thomas Middle School
Miami, Florida
Michael Paloger

Wooddale Elementary Middle School
Memphis, Tennessee
Velma Quinn Hodges
Jacqueline Marie Hunt

Yahara Elementary
Stoughton, Wisconsin
Mary Bennett
Kevin Wright

Site Coordinators

Mary L. Delagardelle—Ames Community Schools, Ames, Iowa

Dr. Hector Hirigoyen—Miami, Florida

Audrey Jackson—Parkway School District, Chesterfield, Missouri

Jorge M. López—Puerto Rico

Susan Militello—Memphis, Tennessee

Carol Pudlin—Culver City, California

Reviewers and Consultants

Michael N. Bleicher
Professor of Mathematics
University of Wisconsin–Madison
Madison, WI

Diane J. Briars
Mathematics Specialist
Pittsburgh Public Schools
Pittsburgh, PA

Donald Chambers
Director of Dissemination
University of Wisconsin–Madison
Madison, WI

Don W. Collins
Assistant Professor of Mathematics Education
Western Kentucky University
Bowling Green, KY

Joan Elder
Mathematics Consultant
Los Angeles Unified School District
Los Angeles, CA

Elizabeth Fennema
Professor of Curriculum and Instruction
University of Wisconsin–Madison
Madison, WI

Nancy N. Gates
University of Memphis
Memphis, TN

Jane Donnelly Gawronski
Superintendent
Escondido Union High School
Escondido, CA

M. Elizabeth Graue
Assistant Professor of Curriculum and Instruction
University of Wisconsin–Madison
Madison, WI

Jodean E. Grunow
Consultant
Wisconsin Department of Public Instruction
Madison, WI

John G. Harvey
Professor of Mathematics and Curriculum & Instruction
University of Wisconsin–Madison
Madison, WI

Simon Hellerstein
Professor of Mathematics
University of Wisconsin–Madison
Madison, WI

Elaine J. Hutchinson
Senior Lecturer
University of Wisconsin–Stevens Point
Stevens Point, WI

Richard A. Johnson
Professor of Statistics
University of Wisconsin–Madison
Madison, WI

James J. Kaput
Professor of Mathematics
University of Massachusetts–Dartmouth
Dartmouth, MA

Richard Lehrer
Professor of Educational Psychology
University of Wisconsin–Madison
Madison, WI

Richard Lesh
Professor of Mathematics
University of Massachusetts–Dartmouth
Dartmouth, MA

Mary M. Lindquist
Callaway Professor of Mathematics Education
Columbus College
Columbus, GA

Baudilio (Bob) Mora
Coordinator of Mathematics & Instructional Technology
Carrollton-Farmers Branch
Independent School District
Carrollton, TX

Paul Trafton
Professor of Mathematics
University of Northern Iowa
Cedar Falls, IA

Norman L. Webb
Research Scientist
University of Wisconsin–Madison
Madison, WI

Paul H. Williams
Professor of Plant Pathology
University of Wisconsin–Madison
Madison, WI

Linda Dager Wilson
Assistant Professor
University of Delaware
Newark, DE

Robert L. Wilson
Professor of Mathematics
University of Wisconsin–Madison
Madison, WI

Dear Teacher,

Welcome! *Mathematics in Context* is designed to reflect the National Council of Teachers of Mathematics Standards for School Mathematics and to ground mathematical content in a variety of real-world contexts. Rather than relying on you to explain and demonstrate generalized definitions, rules, or algorithms, students investigate questions directly related to a particular context and construct mathematical understanding and meaning from that context.

The curriculum encompasses 10 units per grade level. *Ups and Downs* is designed to be the seventh unit in the algebra strand for grade 7/8, but it also lends itself to independent use—to introduce students to experiences that will enrich their understanding for different ways of growth and that will provide students a base to investigate linear and exponential growth in a more formal way in higher grades.

In addition to the Teacher Guide and Student Books, *Mathematics in Context* offers the following components that will inform and support your teaching:

• *Teacher Resource and Implementation Guide*, which provides an overview of the complete system, including program implementation, philosophy, and rationale

• *Number Tools*, Volumes 1 and 2, which are a series of blackline masters that serve as review sheets or practice pages involving number issues and basic skills

• *News in Numbers*, which is a set of additional activities that can be inserted between or within other units; it includes a number of measurement problems that require estimation.

Thank you for choosing *Mathematics in Context*. We wish you success and inspiration!

Sincerely,

The Mathematics in Context Development Team

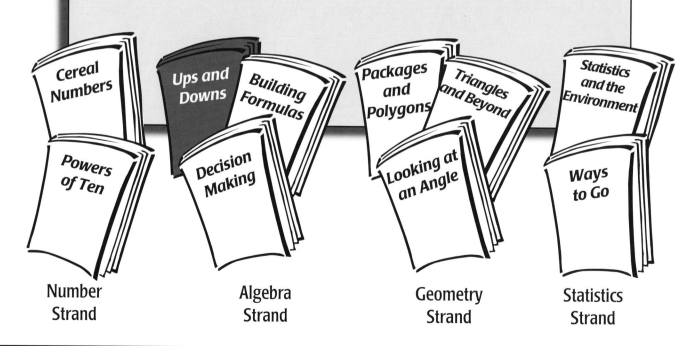

| Number Strand | Algebra Strand | Geometry Strand | Statistics Strand |

Overview

B R I T A N N I C A

**Mathematics
in
Context**

BUS SCHEDULE

OUTDOOR 96.8 °F

INDOOR 77.5 °F

Gino's

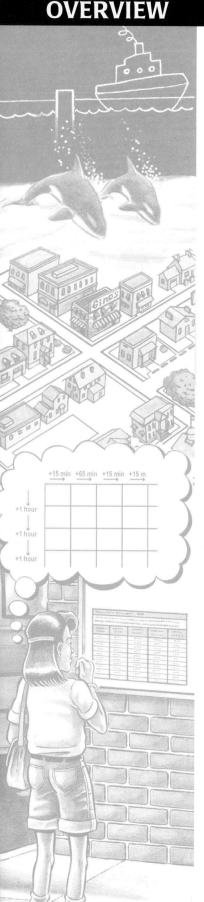

How to Use This Book

This unit is one of 40 for the middle grades. Each unit can be used independently; however, the 40 units are designed to make up a complete, connected curriculum (10 units per grade level). There is a Student Book and a Teacher Guide for each unit.

Each Teacher Guide comprises elements that assist the teacher in the presentation of concepts and in understanding the general direction of the unit and the program as a whole. Becoming familiar with this structure will make using the units easier.

Each Teacher Guide consists of six basic parts:

- Overview
- Student Materials and Teaching Notes
- Assessment Activities and Solutions
- Glossary
- Blackline Masters
- Try This! Solutions

Overview

Before beginning this unit, read the Overview in order to understand the purpose of the unit and to develop strategies for facilitating instruction. The Overview provides helpful information about the unit's focus, pacing, goals, and assessment, as well as explanations about how the unit fits with the rest of the *Mathematics in Context* curriculum.

Student Materials and Teaching Notes

This Teacher Guide contains all of the student pages, (except the Try This! activities), each of which faces a page of solutions, samples of students' work, and hints and comments about how to facilitate instruction. Note: Solutions for the Try This! activities can be found in the back of the Teacher Guide.

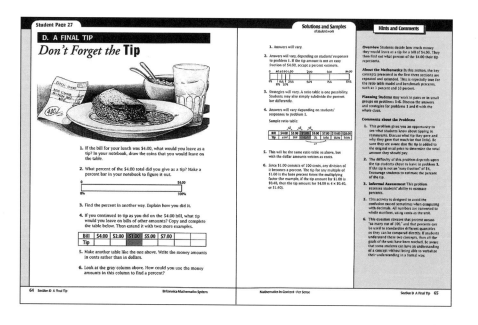

Each section within the unit begins with a two-page spread that describes the work students do, the goals of the section, new vocabulary, and materials needed, as well as providing information about the mathematics in the section and ideas for pacing, planning instruction, homework, and assessment.

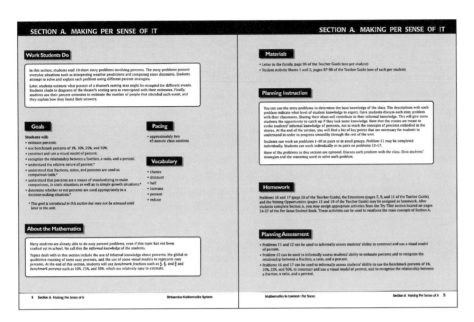

Assessment Activities and Solutions

Information about assessment can be found in several places in this Teacher Guide. General information about assessment is given in the Overview; informal assessment opportunities are identified on the teacher pages that face each student page; and the Assessment Activities section of this guide provides formal assessment opportunities.

Glossary

The Glossary defines all vocabulary words listed on the Section Opener pages. It includes the mathematical terms that may be new to students, as well as words associated with the contexts introduced in the unit. (Note: The Student Book does not have a glossary. This allows students to construct their own definitions, based on their personal experiences with the unit activities.)

Blackline Masters

At the back of this Teacher Guide are blackline masters for photocopying. The blackline masters include a letter to families (to be sent home with students before beginning the unit), several student activity sheets, and assessment masters.

Try This! Solutions

Also included in the back of this Teacher Guide are the solutions to several Try This! activities—one related to each section of the unit—that can be used to reinforce the unit's main concepts. The Try This! activities are located in the back of the Student Book.

Unit Focus

In *Ups and Downs,* students are informally introduced to a variety of real-life situations involving growth patterns. Throughout the unit, students learn how to describe these patterns of increase and decrease using tables of values, graphs, and formulas expressed as word formulas, arrow strings, or with letter symbols. The mathematics in Section A focuses on examining patterns of increase and decrease in tables and graphs. Graphs are analyzed and interpreted in greater depth here than in previous units, as students are challenged to find the differences in growth in a graph and relate tables of values to various graphs of straight lines and curves. In later sections, periodic functions, linear functions, and exponential growth and decay are introduced. Students use both recursive and direct formulas to describe linear growth patterns, while exponential growth is described using only recursive formulas. In each growth pattern context, the relationship between a formula's table of values and its graph is investigated and analyzed. In this unit, students are also challenged to give the meanings of the variables and numbers in specific formulas and explain the origin of some formulas.

Mathematical Content

- using functions of time (linear, nonlinear, increasing, decreasing, periodic, exponential)
- using graphs and tables
- understanding recursive and direct formulas

Prior Knowledge

This unit assumes that students have an understanding of the following:

- measuring heights, lengths, and distances in metric units;
- making a graph from data in a table (from the unit *Tracking Graphs*);
- describing relationships with word formulas (from the unit *Expressions and Formulas*).

In addition, students should have some experience with finding both local and global information on a graph (from the unit *Tracking Graphs*). Students should also have experience in working with symbols in formulas (from the units *Expressions and Formulas* and *Comparing Quantities*). Experiences with the concept of a factor (from the unit *Ratios and Rates*) may be helpful to develop a better understanding of the concept of a growth factor.

Planning and Preparation

Pacing: 19 days

Section	Work Students Do	Pacing*	Materials
A. Differences in Growth	■ investigate increase and decrease in tables and graphs ■ investigate patterns of increase in tables and graphs	6 days	■ Letter to the Family (one per student) ■ Student Activity Sheets 1–4 (one of each per student) ■ centimeter rulers (one per student) ■ paper strips (one per group) ■ measuring tape (one per group) ■ scissors (one pair per student) ■ See page 5 of the Teacher Guide for a complete list of materials and quantities needed.
B. Cycles	■ investigate periodicity in graphs and learn to identify the most important features of periodic graphs (periods and cycles)	3 days	■ Student Activity Sheets 5 and 6 (one of each per student) ■ graph paper (four sheets per student) ■ See page 47 of the Teacher Guide for a list of optional materials and quantities needed.
C. Linear Patterns	■ identify linear growth in tables and graphs ■ make recursive and direct formulas to describe linear growth	5 days	■ Student Activity Sheets 7 and 8 (one of each per student) ■ compasses (one per student) ■ calculators (one per student) ■ graph paper (one sheet per student) ■ colored pencils (one box per student) ■ See page 65 of the Teacher Guide for a list of optional materials and quantities needed.
D. Faster and Faster	■ investigate exponential growth (repeated doubling) in tables and graphs ■ understand and use the concept of growth factor	2 days	■ Student Activity Sheets 9 and 10 (one of each per student) ■ graph paper (one sheet per student) ■ colored pencils (one box per group) ■ calculators (one per student) ■ centimeter rulers (one per student)
E. Half-Lives	■ investigate exponential decay (repeated halving) in tables and graphs	3 days	■ graph paper (two sheets per student) ■ calculators (one per student) ■ millimeter graph paper (one sheet per student)

* One day is approximately equivalent to one 45-minute class session.

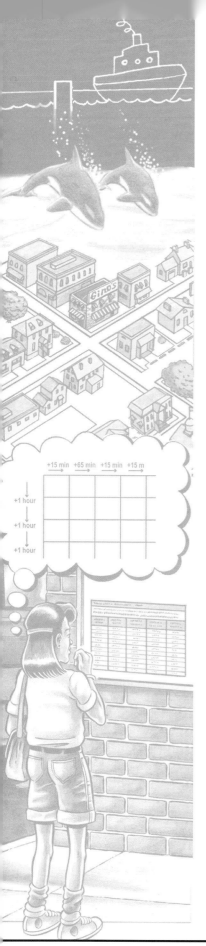

Preparation

In the *Teacher Resource and Implementation Guide* is an extensive description of the philosophy underlying both the content and the pedagogy of the *Mathematics in Context* curriculum. Suggestions for preparation are also given in the Hints and Comments columns of this Teacher Guide. You may want to consider the following:

• Work through the unit before teaching it. If possible, take on the role of the student and discuss your strategies with other teachers.

• Use the overhead projector for student demonstrations, particularly with overhead transparencies of the student activity sheets and any manipulatives used in the unit.

• Invite students to use drawings and examples to illustrate and clarify their answers.

• Allow students to work at different levels of sophistication. Some students may need concrete materials, while others can work at a more abstract level.

• Provide opportunities and support for students to share their strategies, which often differ. This allows students to take part in class discussions and introduces them to alternative ways to think about the mathematics in the unit.

• In some cases, it may be necessary to read the problems to students or to pair students to facilitate their understanding of the printed materials.

• A list of the materials needed for this unit is in the chart on page xiii.

• Try to follow the recommended pacing chart on page xiii. You can easily spend much more time on this unit than the number of class periods indicated. Bear in mind, however, that many of the topics introduced in this unit will be revisited and covered more thoroughly in other *Mathematics in Context* units.

Resources

For Teachers

Books and Magazines
• *Mathematics Assessment: Myths, Models, Good Questions, and Practical Suggestions,* edited by Jean Kerr Stenmark (Reston, Virginia: The National Council of Teachers of Mathematics, Inc., 1991)

Assessment

Planning Assessment

In keeping with the NCTM Assessment Standards, valid assessment should be based on evidence drawn from several sources. (See the full discussion of assessment philosophies in the *Teacher Resource and Implementation Guide.*) An assessment plan for this unit may draw from the following sources:

- Observations—look, listen, and record observable behavior.

- Interactive Responses—in a teacher-facilitated situation, note how students respond, clarify, revise, and extend their thinking.

- Products—look for the quality of thought evident in student projects, test answers, worksheet solutions, or writings.

These categories are not meant to be mutually exclusive. In fact, observation is a key part in assessing interactive responses and also key to understanding the end results of projects and writings.

Ongoing Assessment Opportunities

- ### Problems within Sections
  To evaluate ongoing progress, *Mathematics in Context* identifies informal assessment opportunities and the goals that these particular problems assess throughout the Teacher Guide. There are also indications as to what you might expect from your students.

- ### Section Summary Questions
  The summary questions at the end of each section are vehicles for informal assessment (see Teacher Guide pages 42, 62, 88, 102, and 118).

End-of-Unit Assessment Opportunities

In the back of this Teacher Guide, there are eight assessments that, when combined, form a two-class period end-of-unit assessment. For a more detailed description of the assessment activity, see the Assessment Overview (Teacher Guide pages 120 and 121).

You may also wish to design your own culminating project or let students create one that will tell you what they consider important in the unit. For more assessment ideas, refer to the chart on pages xvi and xvii.

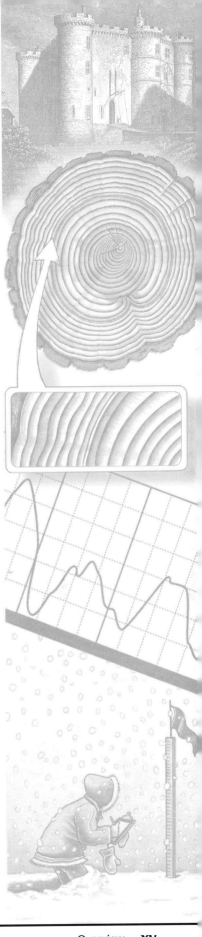

Goals and Assessment

In the *Mathematics in Context* curriculum, unit goals, categorized according to cognitive procedures, relate to the strand goals and to the NCTM Curriculum and Evaluation Standards. Additional information about these goals is found in the *Teacher Resource and Implementation Guide.* The *Mathematics in Context* curriculum is designed to help students develop their abilities so that they can perform with understanding in each of the categories listed below. It is important to note that the attainment of goals in one category is not a prerequisite to attaining those in another category. In fact, students should progress simultaneously toward several goals in different categories.

	Goal	Ongoing Assessment Opportunities	End-of-Unit Assessment Opportunities
Conceptual and Procedural Knowledge	**1.** use information about increase and/or decrease to create line graphs	**Section A** p. 20, #14 **Section C** p. 80, #26	Dots, Distances, and Speed, p. 154 Taxi!, p. 157
	2. identify and describe patterns of increase and/or decrease from a table or graph	**Section A** p. 16, #9 p. 20, #15, #16 p. 30, #29 p. 36, #38 p. 44, #43–#45 **Section D** p. 96, #7 **Section E** p. 118, #19	Make Up a Story, p. 153 Dots, Distances, and Speed, p. 154 Taxi!, p. 157 Bacteria in Food, p. 159 Radioactivity, p. 160–161
	3. identify characteristics of periodic graphs	**Section B** p. 56, #11–13 p. 62, #21	Deep Water, p. 155 Lighthouse, p. 156
	4. identify linear patterns in tables and graphs	**Section C** p. 88, #42a	Taxi!, p. 157 Growth, p. 158
	5. understand and use the concept of growth factor	**Section D** p. 96, #6 p. 102, #17, #18 **Section E** p. 110, #10	Growth, p. 158 Bacteria in Food, p. 159 Radioactivity, p. 160–161

	Goal	Ongoing Assessment Opportunities	End-of-Unit Assessment Opportunities
Reasoning, Communicating, Thinking, and Making Connections	**6.** describe linear growth with recursive formulas	**Section C** p. 76, #18 p. 88, #42b, #43b	Growth, p. 158
	7. describe linear growth with direct formulas	**Section C** p. 80, #27 p. 88, #42c, #43c	Taxi!, p. 157 Growth, p. 158
	8. make connections between situation, graph, and table	**Section A** p. 16, #9 p. 30, #29 p. 44, #43–#45 **Section B** p. 56, #11 **Section C** p. 88, #43a	Make Up a Story, p. 153 Dots, Distances, and Speed, p. 154
	9. reason about situations of growth in terms of slope, maximum and minimum, range, decrease, and increase	**Section A** p. 20, #17 p. 30, #29 p. 36, #38 **Section C** p. 80, #25 **Section E** p. 118, #17, #18	Make Up a Story, p. 153 Dots, Distances, and Speed, p. 154 Bacteria in Food, p. 159
	10. identify a growth factor	**Section D** p. 102, #18	Bacteria in Food, p. 159 Radioactivity, p. 160–161

	Goal	Ongoing Assessment Opportunities	End-of-Unit Assessment Opportunities
Modeling, Nonroutine Problem-Solving, Critically Analyzing, and Generalizing	**11.** recognize the power of graphs and/or tables for representing and solving problems	**Section A** p. 20, #17 **Section C** p. 86, #37, #38 **Section D** p. 96, #7	Deep Water, p. 155 Taxi!, p. 157 Bacteria in Food, p. 159
	12. use algebraic models to represent realistic situations	**Section D** p. 102, #18 **Section E** p. 118, #17, #18	Lighthouse, p. 156 Taxi!, p. 157 Radioactivity, p. 160–161

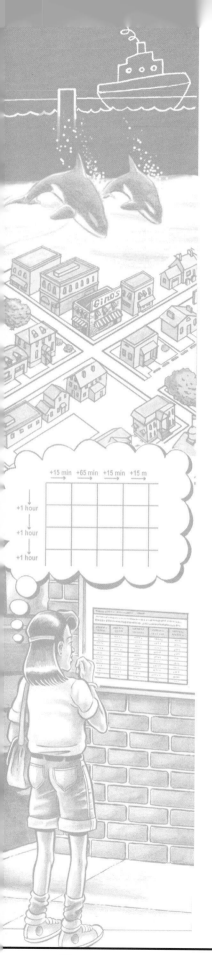

More about Assessment

Scoring and Analyzing Assessment Responses

Students may respond to assessment questions with various levels of mathematical sophistication and elaboration. Each student's response should be considered for the mathematics that it shows, and not judged on whether or not it includes an expected response. Responses to some of the assessment questions may be viewed as either correct or incorrect, but many answers will need flexible judgment by the teacher. Descriptive judgments related to specific goals and partial credit often provide more helpful feedback than percent scores.

Openly communicate your expectations to all students, and report achievement and progress for each student relative to those expectations. When scoring students' responses, try to think about how they are progressing toward the goals of the unit and the strand.

Student Portfolios

Generally, a portfolio is a collection of student-selected pieces that is representative of a student's work. A portfolio may include evaluative comments by you or by the student. See the *Teacher Resource and Implementation Guide* for more ideas on portfolio focus and use.

A comprehensive discussion about the contents, management, and evaluation of portfolios can be found in *Mathematics Assessment: Myths, Models, Good Questions, and Practical Suggestions*, pp. 35–48.

Student Self-Evaluation

Self-evaluation encourages students to reflect on their progress in learning mathematical concepts, their developing abilities to use mathematics, and their dispositions toward mathematics. The following examples illustrate ways to incorporate student self-evaluations as one component of your assessment plan.

- Ask students to comment, in writing, on each piece they have chosen for their portfolios and on the progress they see in the pieces overall.
- Give a writing assignment entitled "What I Know Now about [a math concept] and What I Think about It." This will give you information about each student's disposition toward mathematics as well as his or her knowledge.
- Interview individuals or small groups to elicit what they have learned, what they think is important, and why.

Suggestions for self-inventories can be found in *Mathematics Assessment: Myths, Models, Good Questions, and Practical Suggestions*, pp. 55–58.

Summary Discussion

Discuss specific lessons and activities in the unit—what the student learned from them and what the activities have in common. This can be done in whole-class discussion, in small groups, or in personal interviews.

Connections across the *Mathematics in Context* Curriculum

Ups and Downs is the seventh unit in the algebra strand. The map below shows the complete *Mathematics in Context* curriculum for grade 7/8. This indicates where the unit fits in the algebra strand and in the overall picture.

A detailed description of the units, the strands, and the connections in the *Mathematics in Context* curriculum can be found in the *Teacher Resource and Implementation Guide.*

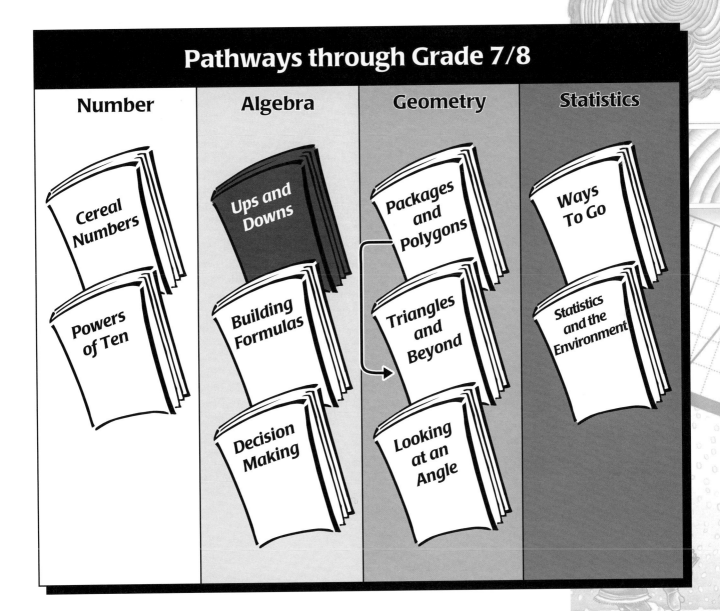

Pathways through Grade 7/8

Number	Algebra	Geometry	Statistics
Cereal Numbers	Ups and Downs	Packages and Polygons	Ways To Go
Powers of Ten	Building Formulas	Triangles and Beyond	Statistics and the Environment
	Decision Making	Looking at an Angle	

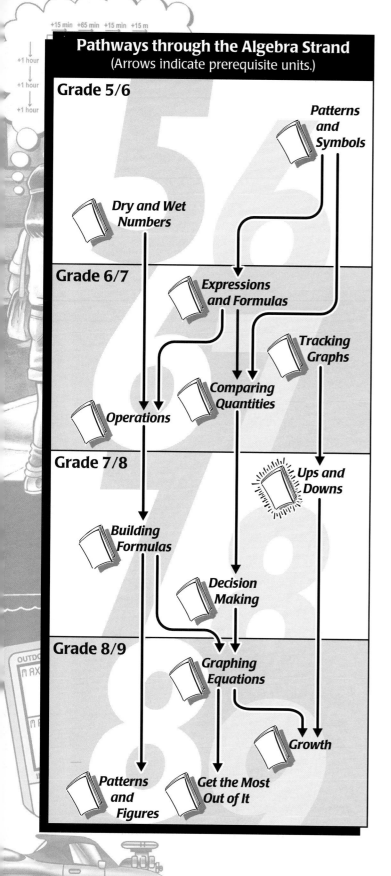

Pathways through the Algebra Strand
(Arrows indicate prerequisite units.)

+15 min +65 min +15 min +15 m

+1 hour
+1 hour
+1 hour

Grade 5/6

Patterns and Symbols

Dry and Wet Numbers

Grade 6/7

Expressions and Formulas

Tracking Graphs

Comparing Quantities

Operations

Grade 7/8

Ups and Downs

Building Formulas

Decision Making

Grade 8/9

Graphing Equations

Growth

Patterns and Figures

Get the Most Out of It

Connections within the Algebra Strand

On the left is a map of the algebra strand; this unit, *Ups and Downs,* is highlighted.

The grade 7/8 unit *Ups and Downs* is the seventh unit in the algebra strand. Refer to the chart on the left for a complete list of all algebra strand units that precede and follow this unit.

Ups and Downs belongs in the algebra substrand dealing with processes, functions that change over time. This topic was informally introduced in the unit *Tracking Graphs. Ups and Downs* builds directly on the concepts and skills developed in that unit, including reading, interpreting, and constructing graphs. Graphs are utilized here to informally investigate processes of change, including linear and quadratic growth patterns, periodic growth, and exponential relationships. These topics are studied more extensively in the unit *Growth.* Graphing is an effective way to represent and communicate about functions of increase and decrease and is an integral component in all three units.

Ups and Downs is also related to the algebra substrand dealing with patterns and regularities. *Expressions and Formulas* is the first of a series of three units that explicitly deals with this topic. In this unit, students explore methods to describe simple relationships and patterns, using word formulas and arrow strings. These concepts are revisited and extended in *Building Formulas.* Formulas are further investigated in *Patterns and Figures,* the third unit of the patterns and regularities substrand.

Ups and Downs is also related to the algebra substrand dealing with restrictions. This concept is first introduced in the unit *Comparing Quantities. Ups and Downs* plays an important role in the study of linear relationships, which leads to the formal study of linear equations expressed in slope-intercept form in the unit *Graphing Equations.*

The Algebra Strand

Grade 5/6

Patterns and Symbols
Using symbols to represent patterns, creating equivalent patterns, and making generalizations from them to find a rule.

Dry and Wet Numbers
Developing the concept of negative numbers, adding with positive and negative numbers, interpreting a drawing using a scale, and making a scale line with positive and negative numbers.

Grade 6/7

Expressions and Formulas
Describing series of calculations using operation strings, generalizing to find a formula, and representing relationships found in tables.

Tracking Graphs
Producing line graphs, reading data from graphs, and learning to look at a graph's essential features rather than details.

Comparing Quantities
Organizing and translating information from stories using symbols, charts, tables, and equations.

Operations
Multiplying, dividing, adding, and subtracting positive and negative numbers and plotting ordered pairs in all four quadrants of the coordinate plane.

Grade 7/8

Building Formulas
Relating tables to algebraic descriptions, applying recursive and direct-rule formulas, relating expressions in equivalent forms, and using squares and square roots to solve area problems.

Ups and Downs
Describing an increasing or decreasing function from a table or graph, determining whether or not a growth function is linear, and recognizing a periodic function.

Decision Making
Representing data with graphs, working with the notion of constraints and with graphing inequalities, and graphing discrete functions.

Powers of Ten
Investigating simple laws for calculating with powers of 10 and investigating very large and very small numbers. (Powers of Ten is also in the number strand.)

Grade 8/9

Graphing Equations
Graphing points and lines in the coordinate plane, solving single-variable linear equations, using inequalities to describe restricted regions on a graph, and learning the structure of the equation of a line.

Patterns and Figures
Recognizing regularity in patterns; expressing generalities; and exploring progressions, rectangular and triangular numbers, and Pascal's triangle.

Growth
Investigating linear, quadratic, cubic, and exponential functions; using recursive and direct formulas; and describing growth with graphs.

Get the Most Out of It
Solving word problems with two unknowns; solving systems of equations; graphing lines, inequalities, and hyperbolas; and working with curved feasible regions.

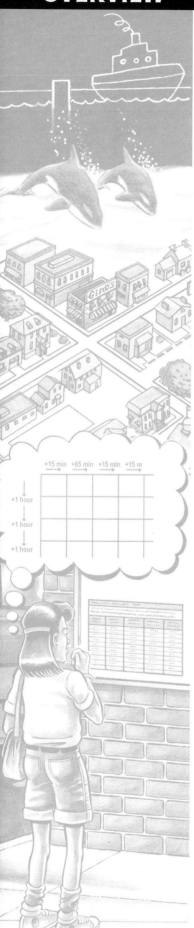

Connections with Other
Mathematics in Context Units

Reading, interpreting, and constructing graphs are important skills in all statistics strand units, beginning with *Picturing Numbers*. Different graphical representations, such as picture graphs, bar graphs, and line graphs, are studied throughout many *Mathematics in Context* units.

The concepts of reduction and enlargement factors, introduced in the unit *Ratios and Rates,* support the development of students' understanding of exponential growth and decay and of the growth factor. Connections are also made with units in the geometry strand. The concept of the steepness of a line (slope) is further investigated in both algebra and geometry units.

The following mathematical topics that are included in the unit *Ups and Downs* are introduced or further developed in other *Mathematics in Context* units.

Prerequisite Topics

Topic	Unit	Grade
scale/ratio	*Figuring All the Angles***	5/6
	*Grasping Sizes**	5/6
	*Ratios and Rates**	6/7
representing information graphically	*Picturing Numbers****	5/6
	Tracking Graphs	6/7
symbolic representation	*Patterns and Symbols*	5/6
	Comparing Quantities	6/7
	Expressions and Formulas	6/7
formulas	*Expressions and Formulas*	6/7
area	*Reallotment***	6/7
	*Made to Measure***	6/7
direct and recursive formulas	*Patterns and Symbols*	5/6

Topics Revisited in Other Units

Topic	Unit	Grade
symbolic representation	*Building Formulas*	7/8
	Graphing Equations	8/9
scale/ratio	*Looking at an Angle***	7/8
	*Going the Distance***	8/9
formulas	*Digging Numbers****	8/9
	*Going the Distance***	8/9
area	*Cereal Numbers**	7/8
slope (as steepness)	*Looking at an Angle*	7/8
	Graphing Equations	8/9
	*Going the Distance***	8/9
direct and recursive formulas	*Building Formulas*	7/8
	*Digging Numbers****	8/9
	*Going the Distance***	8/9
	Growth	8/9
line graphs	*Growth*	8/9
linear functions	all algebra units	
exponential functions	*Building Formulas*	7/8
	Growth	8/9

 * These units in the number strand also help students make connections to ideas about algebra and integers.
 ** These units in the geometry strand also help students make connections to ideas about algebra and integers.
 *** These units in the statistics strand also help students make connections to ideas about algebra and integers.

Student Materials and Teaching Notes

Student Book
Table of Contents

Dear Student,

Welcome to *Ups and Downs*. In this unit, you will look at things that change over time, such as blood pressure or the tides of an ocean. You'll learn to represent these changes using tables, graphs, and formulas.

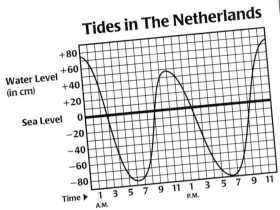

Tides in The Netherlands

Graphs of temperatures and tides go up and down, but some graphs go only upward or only downward, like graphs for tree growth or ice melting.

As you become more familiar with graphs and the changes they represent, you will begin to notice and understand graphs in newspapers, magazines, and advertisements. You will see the advantages of telling a story with a graph.

Sincerely,

The Mathematics in Context Development Team

SECTION A. DIFFERENCES IN GROWTH

Work Students Do

Students use cross sections and core samples of tree trunks and totem poles to determine how much trees grow over specific periods of time. Then students learn how to represent this information on graphs. The concept of growth over time is explored in other contexts, such as a person's height, the diameter of a tree trunk, and the expansion of an oil spill on water. Students also recreate missing information about the growth of sunflowers from a combination of tables, graphs, and reports. Next, students examine the changes in a baby's weight over time as well as the changes in the weight of an ice cube as it melts. These activities demonstrate the usefulness of graphs in illustrating how something increases and decreases over time.

Goals

Students will:

- use information about increase and/or decrease to create line graphs;
- identify and describe patterns of increase and/or decrease from a table or graph;
- make connections between situation, graph, and table;
- reason about situations of growth in terms of slope, maximum and minimum, range, decrease, and increase;
- recognize the power of graphs and/or tables for representing and solving problems.

Pacing

- approximately six 45-minute class sessions

About the Mathematics

In this section, patterns of increase and decrease over time are informally introduced within a variety of contexts. Growth patterns are explored and investigated using diagrams such as tree cross sections, information in tables, and linear (straight line) and quadratic (curved line) graphic representations. The emphasis in this section is on developing students' ability to read, interpret, and make predictions based on information in tables and graphs. Students also make informal observations in comparing tables of information with different graphic representations to see which graph visually displays the data from which table. In Section C, the concept of a linear function is formally defined as a pattern of growth that shows equal differences over equal time periods. Students also learn that the graphs of all linear growth patterns are represented by straight lines.

Materials

- Letter to the Family, page 142 of the Teacher Guide (one per student)
- Student Activity Sheets 1–4, pages 143–146 of the Teacher Guide (one of each per student)
- centimeter rulers, pages 7 and 17 of the Teacher Guide (one per student)
- paper strips and measuring tape, page 7 of the Teacher Guide (one per group of students)
- overhead projector, pages 11, 21, 23, 27, and 35 of the Teacher Guide, optional (one per class)
- transparency of the picture on page 11 of the Teacher Guide, optional (one per class)

- scissors, page 15 of the Teacher Guide (one pair per student)
- millimeter graph paper, pages 17 and 27 of the Teacher Guide (two sheets per student)
- transparency of millimeter graph paper, page 27 of the Teacher Guide, optional (one per class)
- compasses, page 17 of the Teacher Guide (one per student)
- transparency of Student Activity Sheet 2, page 144 of the Teacher Guide, optional (one per class)
- transparency of Student Activity Sheet 3, page 145 of the Teacher Guide, optional (one per class)
- transparencies of the charts on Student Book page 15, page 35 of the Teacher Guide, optional (one per class)
- graph paper, pages 29, 31, and 43 of the Teacher Guide (three sheets per student)
- cooking oil and liquid detergent, page 27 of the Teacher Guide (one per group of students)
- clear, shallow dish, white vinegar, red or blue food coloring, pipet or eye dropper, page 27 of the Teacher Guide (one of each per group of students)

Planning Instruction

You may want to begin with a class discussion about the oldest and largest trees you or your students are familiar with. Some students may have seen or heard of sequoia trees. If students do not mention sequoias in the discussion, you might want to mention them yourself.

Students may work on problems 32 and 33 as a class and work individually on problems 38–41 and 43–45. They may work on the remaining problems in pairs or small groups. Problems 40 and 41 are optional. If time is a concern, you may omit these problems or assign them as homework.

Homework

Problems 6 and 7 (page 14 of the Teacher Guide), 9 (page 16 of the Teacher Guide), 20 (page 22 of the Teacher Guide), and 39–41 (page 40 of the Teacher Guide) may be assigned as homework. Also, the Extension (page 29 of the Teacher Guide) may be assigned as homework. After students complete Section A, you may assign appropriate activities from the Try This! section, located on pages 54–57 of the *Ups and Downs* Student Book. The Try This! activities reinforce the key mathematical concepts introduced in this section.

Planning Assessment

- Problems 9, 15, 16, and 43–45 may be used to informally assess students' ability to identify and describe patterns of increase and/or decrease from a table or graph, and their ability to make connections between situation, graph, and table.
- Problem 14 may be used to informally assess students' ability to use information about increase and/or decrease to create line graphs.
- Problem 17 may be used to informally assess students' ability to reason about situations of growth in terms of slope, maximum and minimum, range, decrease, and increase and to recognize the power of graphs and/or tables for representing and solving problems.
- Problems 29 and 38 may be used to informally assess students' ability to make connections between situation, graph, and table, and to identify and describe patterns of increase and/or decrease from a table or graph. These problems also assess students' ability to reason about situations of growth in terms of slope, maximum and minimum, range, decrease, and increase.

Wooden Graphs

There are giant sequoia trees in Sequoia National Park in California. The largest tree in the park is thought to be between 3,000 and 4,000 years old.

It takes 16 children to reach around the giant sequoia shown in the lower right corner of the picture.

1. Find a way to estimate the circumference and diameter of this tree.

1. Estimates and strategies will vary. Sample response:

When you hold your arms out to your sides, the distance between the fingertips of your left and right hands is the same as your height, measured from head to toe. If you assume that these children are about 1.5 m tall, then the circumference will be about $16 \times 1.5 = 24$ m. Use the relationship between circumference and diameter. The diameter can be found by calculating $24 \text{ m} \div \pi = 7.6$ meters.

Materials centimeter rulers (one per student); paper strips (one per group of students); measuring tape (one per group of students)

Overview Students estimate the circumference and diameter of a giant sequoia tree shown in the picture on Student Book page 1.

About the Mathematics To solve problem **1,** students can use the experiences and knowledge they have acquired in previous units:

- A person's arm span measures about the same as his or her height. Students may remember this from the unit *Made to Measure* and the historic word for arm span: the *fathom*;

- The diameter is the length of a straight line through the center of a circle. The circumference is the perimeter of a circle. These concepts were introduced in the unit *Reallotment*;

- Sizes of pictured objects can be estimated using the standard sizes of familiar things that appear near the object in question. This concept is introduced in the units *Side Seeing* and *Grasping Sizes*;

- The constancy of the ratio between circumference and diameter is a bit more than three (some students may remember it as π). This is one of the concepts studied in *Reallotment*. The formula for circumference of a circle will be reintroduced in the unit *Going the Distance.*

Planning You may wish to have a short class discussion about sequoia trees, focusing on the diameter of the trunks. You might ask students whether they have seen or heard of these trees. Students may work on problem **1** in pairs or in small groups.

Comments about the Problems

1. If students are having difficulty, you may give them some hints, such as reminding them of the concept of arm span.

Cross Section

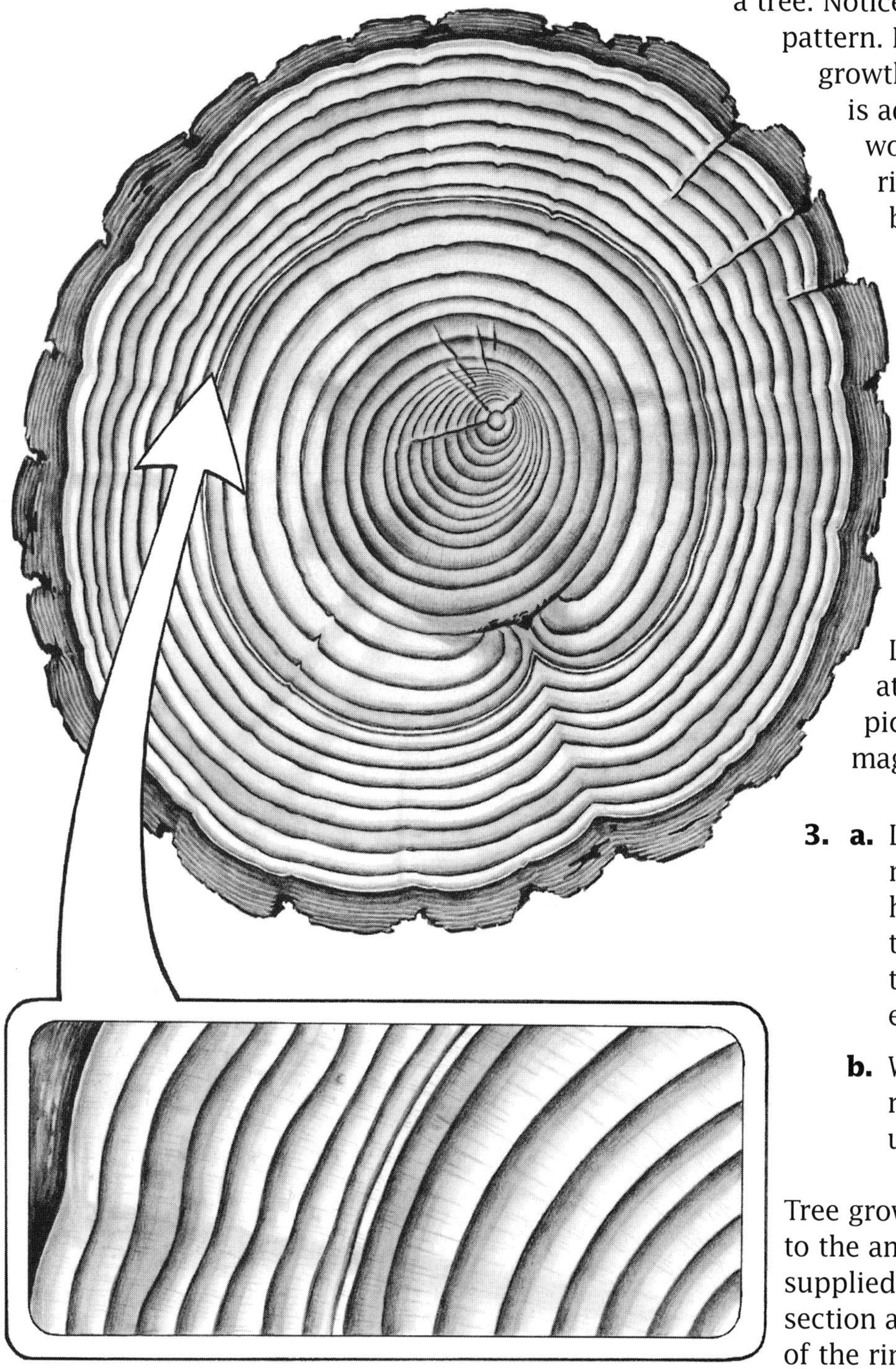

This is a drawing of a cross section of a tree. Notice its distinct ring pattern. During each year of growth, a new layer of cells is added to the older wood. Each layer forms a ring. The distance between the dark rings is how much the tree grew that year.

2. Look at the cross section of a tree on the left. Estimate the age of this tree. How did you find your answer?

Let's take a closer look at the cross section. The picture below shows a magnified portion.

3. a. Looking at the magnified portion, how can you tell that this tree did not grow the same amount each year?

b. What are some possible reasons for the tree's uneven growth?

Tree growth is directly related to the amount of moisture supplied. Look at the cross section again. Notice that one of the rings is very narrow.

Magnified Portion of Cross Section

4. a. What conclusion can you draw about the rainfall during the year that produced the narrow ring?

b. How old was the tree in this year?

2. Answers and strategies will vary. Sample response:

   About 26 or 27 years. I found my answer by counting the rings.

3. **a.** Answers will vary. Some students may say that the rings have different widths.

   b. Answers will vary. Sample responses:
   - lack of water
   - a long, cold winter
   - a spring frost
   - a disease

4. **a.** Answers will vary. Sample response:

      The tree didn't grow very much that year, so perhaps there was not much rain.

   b. The tree was about 19 years old.

Overview Students use a cross section of a tree to estimate the age of the tree and to draw conclusions about the tree's growth.

About the Mathematics Looking at the thickness of the rings of a cross section of a tree shows how much the radius of the tree increased each year. The study of the thickness of the rings will develop students' understanding of different rates of increase, which is one of the main topics of this unit.

Planning Students may work in pairs or small groups on problems **2–4.** After they finish these problems, you might have a short class discussion on the meaning of the thickness of the rings.

Comments about the Problems

2. If students have difficulty seeing that they need to count the rings and not the lines, you might ask students to draw a cross section of a tree with four rings and to ask them to color one of the rings.

4. **b.** It is important that students know that rings grow on the outside rather than from the inside.

CHRIS, HYUN-JIN, AND JOSÉ ARE LOOKING AT THE CROSS SECTIONS OF TWO BIRCH TREES THAT HAD BEEN GROWING NEXT TO EACH OTHER...

I THINK THEY CUT DOWN THE SMALLER TREE FIRST BECAUSE THE OTHER TREE IS BIGGER AND LIVED LONGER.

I THINK THEY CUT DOWN BOTH TREES IN THE SAME YEAR. THE BIGGER ONE WAS JUST PLANTED A FEW YEARS EARLIER.

MAYBE THEY CUT DOWN THE BIG TREE EARLIER. WHO KNOWS?

I JUST REMEMBERED, WE CAN LOOK AT THE RINGS OF THE TREES...

...TO TELL WHAT YEARS THE TREES GREW TOGETHER!

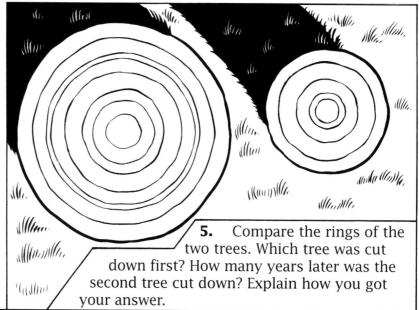

5. Compare the rings of the two trees. Which tree was cut down first? How many years later was the second tree cut down? Explain how you got your answer.

5. The larger tree was cut down one year before the smaller tree. Using the narrow ring to compare both trees, you can see that the smaller tree has a year of growth that the larger tree does not have. The only way this could have happened is if the larger tree had been chopped down, stopping its growth.

Materials transparency of the picture on page 11 of the Teacher Guide, optional (one per class); overhead projector, optional (one per class)

Overview Students solve a problem using the cross sections of two different trees.

Planning Students may work in pairs or in small groups on problem **5.** Discuss students' solutions and strategies for this problem. You might use the overhead projector to show the students how the rings match in the picture on the bottom of this page.

Comments about the Problems

5. If students have difficulties solving this problem, you might suggest that they try to find out which years the trees grew together by looking for matching rings. One way to do this is to trace and cut out the cross sections and match them up as shown below:

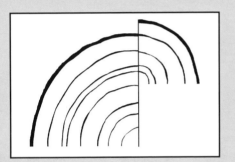

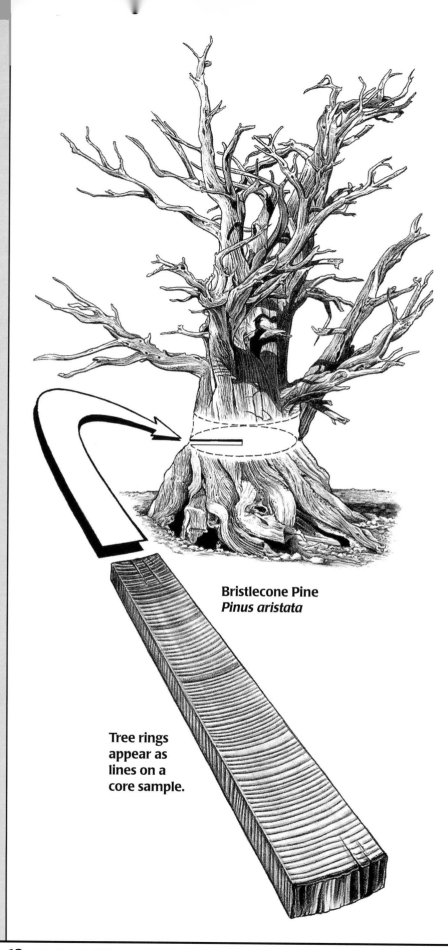

Bristlecone Pine
Pinus aristata

Tree rings appear as lines on a core sample.

The oldest known living tree is a bristlecone pine (*Pinus aristata*) named Methusalah. Methusalah is about 4,700 years old and grows in the White Mountains of California.

It isn't necessary to cut down a tree in order to examine the pattern of rings. Scientists use a technique called *coring* to take a look at the rings of a living tree. They use a special drill to remove a piece of wood from the center of the tree. This piece of wood is about the thickness of a drinking straw and is called a *core sample.* The growth rings show up as lines on the core sample.

By matching the ring patterns from a living tree with those of ancient trees, scientists can create a calendar of tree growth.

Overview Students learn how rings of living trees can be investigated by taking a core sample. They read about the strategy of matching core samples to create a calendar of tree growth. There are no problems on this page for students to solve.

Planning You may want to read and discuss this page as a whole-class activity. Be sure students understand why it is possible to compare the core samples taken from different trees by matching the lines.

The picture below shows how two core samples are matched up. Core sample **B** is from a living tree. Core sample **A** is from a tree that was cut down in the same area. Matching the two samples in this way produces a "calendar" of wood.

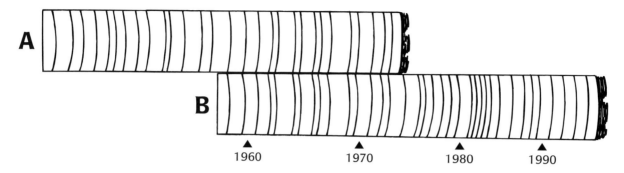

6. In what year was the tree represented by core sample **A** cut down?

The picture below shows a core sample from another tree that was cut down. If you match this one to the other samples, the calendar becomes even longer. Enlarged versions of the three strips can be found on **Student Activity Sheet 1.**

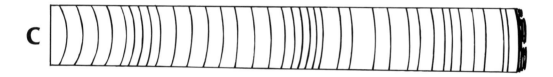

7. What period of time is represented by the three core samples?

Instead of working with the actual core samples or drawings of core samples, scientists transfer the information from the core samples onto a diagram like the one on the right.

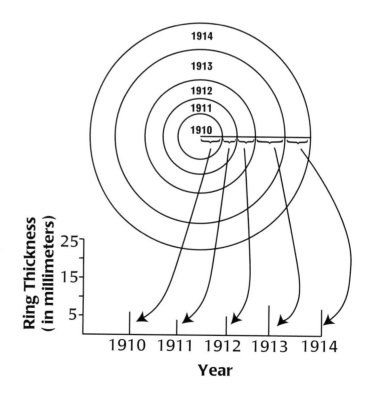

6. 1973

7. The period of time represented by the samples is from 1919 to 1994.

This is the way the patterns match (note that each dot marks the beginning of a new decade):

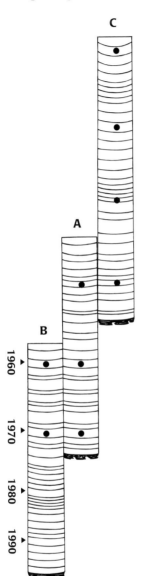

Materials Student Activity Sheet 1 (one per student); scissors (one per student)

Overview Students match three core samples in order to find out what period of time the calendar of wood represents.

About the Mathematics There is a similarity between the diagram of a tree's core sample, shown on the opposite page, and the snapshot diagram of a race, as seen on page 1 of the *Tracking Graphs* Student Book. Both diagrams show a pattern of increase. The main difference is that the core sample diagram shows the growth of the radius in equal time intervals (one year), while the snapshot diagram shows the time it takes various racers to travel equal distances. If the snapshots were taken at equal time intervals, the two pictures would essentially illustrate the same concept. In other words, the differences between the lines and the differences between the plotted points would show the rate of the change in distance traveled over equal time periods, that is, the rate of growth (trees) or the rate of movement (racers).

Planning Students may work on problems **6** and **7** in pairs or small groups. You may want to assign these problems as homework. After students finish the problems, you may want to discuss them in class. After you discuss problem **7** in class, ask students to describe the connection between the tree rings and the graph that are shown on the bottom of Student Book page 5. This connection is investigated on the next Student Book page.

Comments about the Problems

6–7. Homework These problems may be assigned as homework.

7. Note that students should not cut off the letters A, B, and C, or the years. Students will find it easier to count if they mark every 10 years by putting dots on the samples or by coloring every tenth ring.

Totem Pole

Tracy found a totem pole in the woods behind her house. It had fallen over, so Tracy could see the growth rings on the bottom of the pole. She wondered how old the wood was.

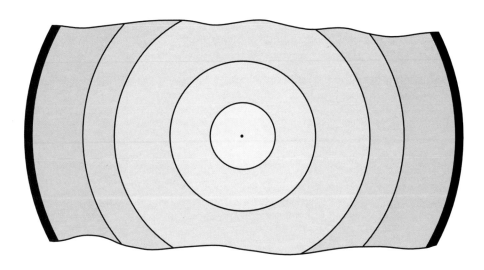

She asked her friend Luis, who studies plants and trees at the university, if he could help her find the age of the wood. He gave her the diagram pictured below, which shows how the cedar trees that were used to make totem poles grew in their area.

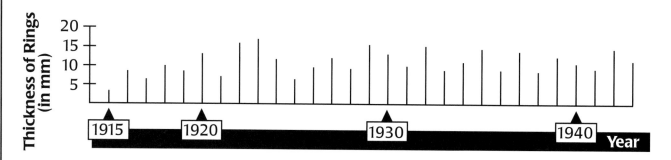

8. **a.** Make a diagram of the rings of the totem pole.

   b. Can you date the totem pole? How old do you think it is?

9. Below is a diagram of the annual rings of a tree. It shows how much the tree grew each year.

   a. Use a ruler and a compass to draw a cross section of this tree.

   b. Write a story that describes how the tree grew.

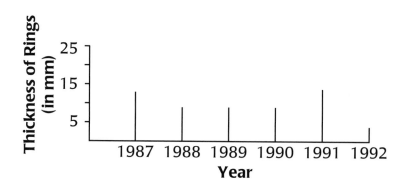

8. a.

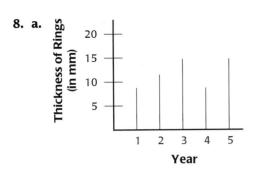

b. The graph matches the period from 1933 to 1937, so the tree was cut down in 1937, and the totem pole was made sometime after that. The totem pole is about 60 to 65 years old.

9. a.

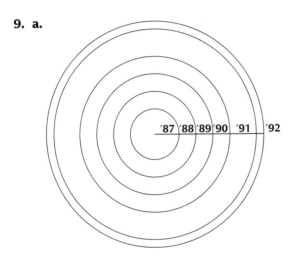

b. Stories will vary. Sample stories:

First it was planted. It got plenty of sun, air, and water and it grew a lot. The second year it grew but not as much as the first year. The third and fourth years, the tree grew about the same as the second year. The fifth year, the tree grew about the same as the first year. The sixth year the tree either had a disease or did not get enough sun or water.

I planted this tree in 1987. For the first year, I watered it a lot and cared for it. It was a very pretty tree and grew a lot in the first year. The second, third, and fourth years I got really bored with it and stopped watering it. It didn't grow much those years. Maybe it grew a couple of inches, but that's all. In 1991, I decided that my tree was very special and started to water it more. It really grew that year. But the next year, I got too busy to water it and it grew very little.

Materials millimeter graph paper, optional (one sheet per student); compasses (one per student); centimeter rulers (one per student)

Overview Students use a diagram to date a wooden totem pole. Then they draw a picture of a cross section of a tree using the information in a diagram of its annual rings, and they write a story about the growth of the tree.

About the Mathematics A diagram that shows the annual increase in the radius of a tree can be used to reconstruct the cross section of the tree. The diagram also can be used to construct a graph that shows the total radius of the tree over time. Note that if you want to have a graph that shows the growth of the diameter over time, you can double the measurements of the radius to get the diameter.

Planning Students may work in pairs or in small groups on problems **8** and **9.** You may use problem **9** as informal assessment or assign it as homework.

Comments about the Problems

8. a. If students are having difficulties, you may refer them to the drawing in problem **7** to see how a diagram can be constructed from a cross section. Observe students measuring the thickness of the rings in millimeters. Students should know that if they want to measure the thickness of the rings precisely, they have to measure along the diameter of the circle. Students may want to use graph paper for drawing the diagram.

b. Wood is usually allowed to dry a few years before it is carved. So the totem could have been made as early as 1937 but probably was made sometime after that year.

9. Informal Assessment This problem assesses students' ability to identify and describe patterns of increase and/or decrease from a table or graph, and to make connections between situation, graph, and table. You may want to assign this problem as homework.

If students are having difficulty drawing circles or measuring distances with a compass, you may want to demonstrate the proper techniques of drawing and measuring with compasses on the board.

Growing Up

On Marsha's birthday, her father marked her height on her bedroom door in centimeters. He did this every year from her first birthday until she was 19 years old.

10. There are only 16 marks. Can you explain this?

11. At what ages did Marsha grow the most?

12. How old was Marsha when her growth slowed dramatically?

13. Where would you put a mark to show Marsha's length at birth?

10. She stopped growing at age 16.

11. She grew the most between three and four years of age, between one and two years of age, and between 11 and 12.

12. Answers will vary. Some students may say age 12, because her growth slowed when she was 12 (the growth from age 12 to age 13 is less than age 11 to age 12). Other students may say age 14, because she grew half the amount of the year before, and for the first time the marks are very close.

13. Answers will vary. A good estimate would be about halfway between the floor and the first mark. Some students may use their own birth length as an estimate. Most newborns average 55 centimeters, or 21 inches, in length.

Overview Students interpret marks that are put on a door every year to indicate the height of a person.

About the Mathematics The marks on the door are similar to the core sample of a tree: the distances between the lines show the yearly increase, or the rate of growth per year.

Planning Students may work on problems **10–13** in pairs or small groups. Be sure to check students' solutions for these problems.

Comments about the Problems

13. One way to think about this problem is to use the doorknob as a point of reference. A doorknob is about 1 meter from the floor. Imagine that a newborn baby would be about half that length, or 50 centimeters. So you would put a mark halfway between the doorknob and the floor.

14. Use **Student Activity Sheet 2** to draw a graph of Marsha's growth. The vertical axis has the same scale as the door, so you can use the marks on the door to get the vertical coordinates.

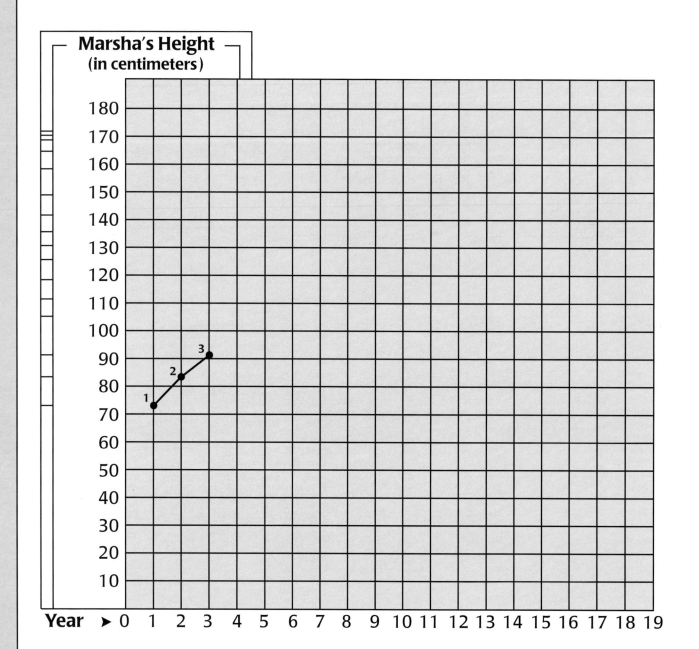

Marsha's Height
(in centimeters)

Year ▸ 0 1 2 3 4 5 6 7 8 9 10 11 12 13 14 15 16 17 18 19

15. When Marsha reached a certain age, her growth slowed down. How does the graph show this?

16. How does the graph show which year she had her biggest growth spurt?

17. Use the graph to estimate her length at birth. Did you get the same length as in problem **13?**

14. Sample graph:

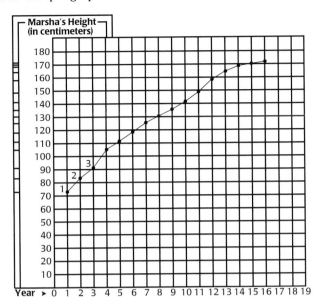

15. The graph levels off starting at age 12. Before age 12, the graph is rising steadily.

16. The biggest growth spurt occurs where the graph is rising most sharply, which is between ages three and four.

17. Answers will vary. Some students might say that if you extend the curve of the graph, it will meet the vertical axis at about 60 centimeters. Other students might say 50 centimeters because they know babies grow quickly during their first year, so the graph should be steep between years 0 and 1.

Materials Student Activity Sheet 2 (one per student); transparency of Student Activity Sheet 2, optional (one per class); overhead projector, optional (one per class)

Overview Students use the marks on the door to create a graph that shows a person's height over time.

About the Mathematics Line graphs are created to show how a person's height grows over time. This graph can be used to make connections between the differences in growth. For example, a larger distance between two marks on the door indicates a growth spurt during that period, and the graph for that period is, therefore, steeper. The steepness of a curve over a given period of time lays the foundation for the study of rate of change in calculus.

Planning Students may work on problems **14–17** in pairs or in small groups. Problems **14–17** may be used as informal assessment. You might discuss these problems focusing on the relationship between the positions of the marks on the door and the steepness of the graph. Problem **14** is critical because it is the first time in the unit that students construct a graph. You may draw the graph of Marsha's growth on a transparency.

Comments about the Problems

14. **Informal Assessment** This problem assesses students' ability to use information about increase and/or decrease to create line graphs.

15–16. **Informal Assessment** These problems assess students' ability to identify and describe patterns of increase and/or decrease from a table or graph.

17. **Informal Assessment** This problem assesses students' ability to reason about situations of growth in terms of slope, maximum and minimum, range, decrease, and increase, and to recognize the power of graphs and/or tables for representing and solving problems.

Students should realize that it is easier to project a growth pattern forward or backward in time when you have a graph. To answer the second part of the question, students should mark their answer to problem **13** on the scale to the left of their graph to compare answers.

Dean is Marsha's brother. The table on the right shows his growth.

18. Draw a graph of Dean's height on **Student Activity Sheet 3.**

Age (in years)	Height (in centimeters)
1	80
2	95
3	103
4	109
5	114
6	118
7	124
8	130
9	138
10	144
11	150
12	156
13	161
14	170
15	176
16	181
17	185
18	187

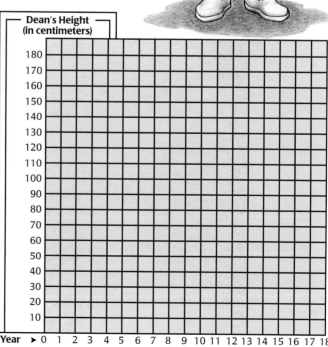

Dean was born a couple of days after Marsha's second birthday. Put Dean's graph (**Student Activity Sheet 3**) on top of Marsha's graph (**Student Activity Sheet 2**) and hold them both up to the light or put them on a window.

19. a. At what age did Dean have his biggest growth spurt?

   b. Who grew faster at that age, Dean or Marsha? How do you know?

20. Make up four questions about these two graphs and the ways in which Marsha and Dean grew. Then answer your own questions.

Dean's Height (in centimeters)

180
170
160
150
140
130
120
110
100
90
80
70
60
50
40
30
20
10

Year ➤ 0 1 2 3 4 5 6 7 8 9 10 11 12 13 14 15 16 17 18

18.

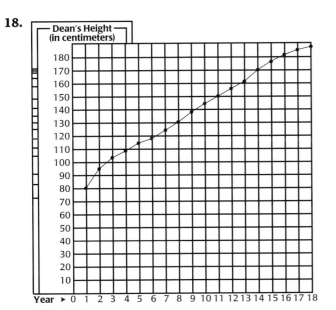

19. a. between his first and second birthdays

   b. Dean grew faster. Explanations will vary. Sample response:

   Dean grew 15 centimeters that year. Marsha's growth spurt was also the biggest between her first and second birthdays, but she only grew about 10 centimeters.

20. Questions will vary. Sample student questions:

   • Who was taller at birth?

   Most students will answer that Marsha was taller at birth, about 63 cm, while Dean was about 60 cm or maybe less.

   • Compare their heights when they both were six years of age. Who was taller at that age? Explain your answer.

   Some students may say that when the graphs are put together with the sixth birthdays lined up, they can see that Marsha is a little taller because the graph is just a little bit higher. Other students may read Marsha's height from the graph (about 119 cm) and Dean's height from the table (118 cm) and then draw the conclusion.

   • How much taller than Marsha do you think Dean will be as an adult?

   Answers will vary. Some students may say that Marsha will not grow beyond her height at age 16—around 172 cm—and that Dean may reach a height of 190 cm.

Materials Student Activity Sheet 3 (one per student); transparency of Student Activity Sheet 3, optional (one per class); overhead projector, optional (one per class)

Overview Students use a table to create a graph showing the growth of Marsha's brother. Then they use both graphs to solve problems.

About the Mathematics Students compare growth rates by comparing one-year periods as shown on two line graphs. This is a focus on steepness. These investigations of differences in growth rates can be considered an informal introduction to calculus. The average growth speed for a given period can be calculated by taking the ratio of the difference in growth to the difference in time.

Planning Students may work in pairs or small groups on problems **18–20.** Problem **20** may be assigned as homework.

Comments about the Problems

20. Homework This problem may be assigned as homework.

   After students complete problem **20,** you might ask a student to read one of the questions he or she made up and give students time to find an answer. Then you may ask a volunteer to show the class how he or she got the answer. It may be useful to have a transparency of Marsha's graph and one of Dean's available.

   The transparencies of the graphs for Dean and Marsha can be used as follows:

   • To compare the heights at a certain age, overlay one transparency on the other with the ages matching.

   • To compare the heights in a certain calendar year, Dean's transparency can be moved to the right by two years.

   • To compare the growth rates at a certain age, overlay one transparency on the other with the ages matching and then move one transparency up or down until the graphs meet each other. Then it is easy to see which is steeper.

The Tree Nursery

Mr. Akimo owns a tree nursery. He measures the diameter of the trunks with a caliper to check their growth. One spring, he selected two different kinds of trees to study.

The two trees were of different species, but they each had trunks that measured two centimeters in diameter. For the next two springs he measured the diameters of both tree trunks. The results are shown in the table below.

Diameters (in centimeters)

	First Measurement	Second Measurement	Third Measurement
Tree 1	2.0	3.0	4.9
Tree 2	2.0	5.5	7.1

21. Which tree will most likely have the larger diameter when Mr. Akimo measures them again next spring? Explain how you got your answer.

21. The first tree will have the larger diameter. Explanations will vary. Sample responses:

The first tree is growing by an increasing amount every year, while the second tree is growing by a decreasing amount every year. This can be shown in the table using arrows with numbers that represent the differences:

Diameters (in centimeters)

	First Measurement	Second Measurement	Third Measurement
Tree 1	2.0	3.0	4.9

+ 1.0 cm   + 1.9 cm

Diameters (in centimeters)

	First Measurement	Second Measurement	Third Measurement
Tree 2	2.0	5.5	7.1

+ 3.5 cm   + 1.6 cm

The first tree might grow as much as 4 cm, putting it at 8.9 cm. The second tree will probably grow less than 1 cm, putting it at about 8 cm. This can be shown by drawing a graph.

Growth of Trees 1 and 2

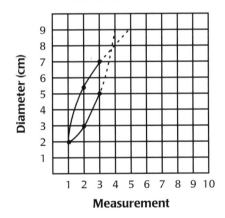

Overview Students investigate a series of measurements of the diameters of growing trees and predict the measurements for the following year.

Planning Students may work in pairs or in small groups on problem **21.** You may want to have students discuss their conclusions and strategies in their groups and then ask them to report their conclusions in class.

Comments about the Problems

21. It is important that students realize that although they can find a growth pattern, it is not certain that the same pattern will continue. This is also a good time to discuss the advantages and disadvantages of the different representations: the table, graph, and growth story (which can be told in words or shown in a cross section drawing).

Activity

Oil Disaster: An Experiment

An oil tanker named the *Exxon Valdez* ran aground in Prince William Sound, off the coast of Alaska, in March of 1989. Over 40 million liters of oil spilled into the sound from the gashed hull of the ship. The oil covered almost 2,000 kilometers of coastline.

Floating barriers called *booms*, like the one in the photo, are sometimes used to contain spills at sea. The boom helps hold the oil in one area until it can be captured with pumps or skimmers. Chemical dispersants are sometimes used to break down the oil, and sometimes it is possible to burn the oil off of the surface of the ocean. These techniques must be used quickly, before wind and waves make the oil slick spread.

For this activity, you will create your own oil spill in order to find out how the area of the slick is related to the amount spilled.

You will need a clear, shallow dish (like a glass pie plate), white vinegar, food coloring, cooking oil, a pipet or eye dropper, a piece of graph paper (with a grid smaller than one centimeter), and some liquid detergent.

First add vinegar to the shallow dish so that the bottom is covered (about one or two centimeters deep is enough). Stir in about eight drops of food coloring so that your ocean will be dark enough to see the oil slick.

Put the dish on the graph paper and let your ocean calm down for a minute or two.

22. Use the pipet to drop two drops of the cooking oil onto the center of your ocean. Use the graph paper under the ocean or a ruler to measure the diameter of your oil slick.

22. Answers will vary, depending on the volume of the drops.

Materials millimeter graph paper (one sheet per student); transparency of millimeter graph paper, optional (one per class); clear, shallow dish, white vinegar, red or blue food coloring, pipet or eye dropper, cooking oil, liquid detergent (one of each per group); overhead projector, optional (one per class)

Overview In this activity, students investigate how an oil spill grows over time.

Planning You might introduce this activity with a class discussion about oil spills, using the text on Student Book page 26. You can, for example, ask students about the size of 40 million liters of oil. Is this the volume of one classroom? More? How long is a coastline of 2,000 kilometers? Then focus the discussion on the purpose of this activity. Students are going to investigate how the size of the oil spill is related to the amount spilled. You may ask: *Will double the amount of oil result in an oil spill that is twice as large?* Students will find a correct answer when they carry out the activity.

Students may work in pairs or in small groups on problem **22.** If the organization of this activity is a problem, you may decide to do this together as a whole-class activity. You can put the shallow dish on the overhead projector on top of a transparency of graph paper with a millimeter grid.

Comments about the Problems

22. Vinegar seems to work better than water. Any residual film of detergent soap can interfere with the oil in water. Students should measure the diameter in millimeters.

Interdisciplinary Connection Certain topics in this activity can be used to make connections with branches of science, like biology and physics. Students may be able to find answers to the following questions:

• Where does oil come from?

• For what purposes can oil be used?

• How is oil refined?

• If oil is spilled in the ocean, what are the effects on living things? What problems do birds have when they get covered by oil?

23. Add two more drops of oil to the center of your oil slick and measure its diameter. Make a table to record your results.

24. Look at the pattern after six drops of oil have been added. What do you predict the diameter of the slick will be for 12 drops of oil? 24 drops?

25. Continue your experiment until at least 30 drops of cooking oil have been added. Fill in your table. Were your predictions correct?

26. Make a graph of your results.

27. If you continued your experiment to 40 drops, how large would you expect the oil slick to be?

28. One way that marine animals are cleaned after an oil spill is with a detergent. What do you think detergent does to oil? Add one drop of liquid detergent to your oil slick. What happened?

23. Tables will vary. See answer to problem **25** for sample table.

24. Answers will vary. Some students may predict that the diameter of the oil spill of 12 drops should be less than two times the diameter of the oil spill made by six drops. Other students may predict that the size of the oil spill will increase at the same rate that it did when it went from containing four drops to containing six drops.

25. Tables will vary. Sample table:

Drops	Diameter (in mm)
2	4
4	5
6	6
8	6.5
10	almost 7
12	7
14	7.5
16	8
18	8.25
20	8.5
22	almost 9
24	9
26	about 9.2
28	about 9.8
30	about 10

26. Graphs will vary. Sample graph:

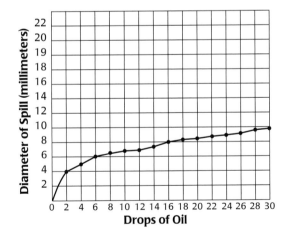

27. Answers will vary. Most students' predictions will reflect their experimental results, showing that the growth in the size of the oil spill does not increase linearly as more oil is added.

28. Answers will vary. Some students may say that the detergent breaks up the oil.

Materials graph paper (one sheet per student)

Overview Students continue with the oil spill activity.

Planning Students may work on problems **23–28** in pairs or in small groups. When students have completed the activity, you might have a class discussion. You may ask a group of students to present their results to the whole class and to evaluate their predictions about how their oil spill would grow. Then focus the discussion on the importance of the use of models. Explain that you can investigate statements or predictions using a model of the real situation. You might emphasize that tables and graphs are important because they can help in getting a better view of the results of measurements and the relationships between them.

Comments about the Problems

23. Students should make measurements in millimeters.

24. Some students may not understand that they have to continue to add two drops until they have an oil spill of six drops.

If students are having difficulty, you may explain to them that they have to look at the way the diameter increases. Note that the differences between the diameters do not show a real pattern, but they do show that as each oil drop is added, the diameter increases at a slower and slower rate.

27. Students may use the results in the table or graph.

Ask students how they made their predictions, whether they prefer to use a table or a graph, and why.

Extension You may ask some students to investigate a more difficult problem: how the area of your oil slick increases. To answer this problem, students have to assume that the oil slick has the shape of a circle. They can use the diameter to find the radius and then calculate the area of the oil slick. They can use an estimation for the area: *area equals about radius × radius × 3*, or they can use the formula *area = π × radius × radius*.

Sunflowers

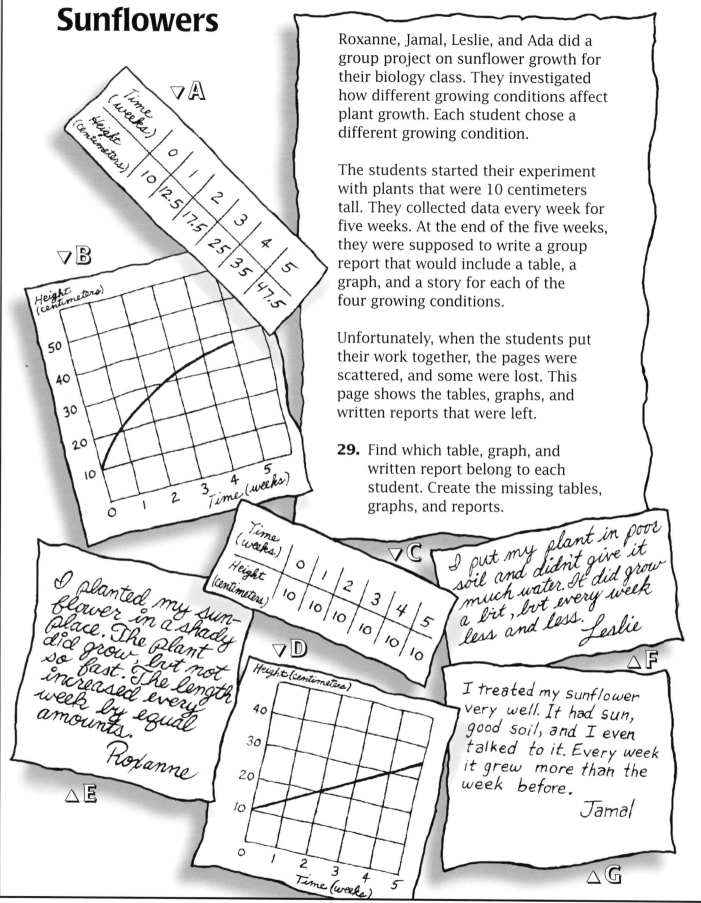

Roxanne, Jamal, Leslie, and Ada did a group project on sunflower growth for their biology class. They investigated how different growing conditions affect plant growth. Each student chose a different growing condition.

The students started their experiment with plants that were 10 centimeters tall. They collected data every week for five weeks. At the end of the five weeks, they were supposed to write a group report that would include a table, a graph, and a story for each of the four growing conditions.

Unfortunately, when the students put their work together, the pages were scattered, and some were lost. This page shows the tables, graphs, and written reports that were left.

29. Find which table, graph, and written report belong to each student. Create the missing tables, graphs, and reports.

▽A

Time (weeks)	0	1	2	3	4	5
Height (centimeters)	10	12.5	17.5	25	35	47.5

▽C

Time (weeks)	0	1	2	3	4	5
Height (centimeters)	10	10	10	10	10	10

I put my plant in poor soil and didn't give it much water. It did grow a bit, but every week less and less. Leslie ▽F

I planted my sunflower in a shady place. The plant did grow, but not so fast. The length increased every week by equal amounts. Roxanne △E

I treated my sunflower very well. It had sun, good soil, and I even talked to it. Every week it grew more than the week before. Jamal △G

29. Answers will vary. Sample response:

Roxanne

  missing table:

Time (weeks)	0	1	2	3	4	5
Height (cm)	10	15	20	25	30	35

Jamal

  missing graph:

Leslie

F B missing table:

Time (weeks)	0	1	2	3	4	5
Height (cm)	10	22	27	33	35	37

Ada

C missing graph:

missing story:

> My sunflower didn't grow at all. I kept it in the dark and didn't water it at all.
>
> *Ada*

Overview Students match tables, graphs, and stories, and they recreate the ones that are missing.

About the Mathematics Some students may think that these kinds of tables are always ratio tables. An explanation of how to tell whether or not a table is a ratio table follows. The table shown below is a ratio table.

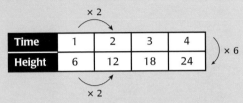

This table is a ratio table because the ratio between time and height is 1:6. In other words, if you double (triple, etc.) the time then you get double (triple, etc.) the height. The graph that corresponds to this table is a straight line (starting at the origin).

The relationship between the numbers in the following table is different:

Time	1	2	3	4
Height	15	20	25	30

This is not a ratio table. The graph is a straight line, but it doesn't start at the origin. Linear growth patterns are investigated more in Section C of this unit.

Planning Students may work in pairs or in small groups on problem **29**. This problem may be used as informal assessment.

Comments about the Problems

29. Informal Assessment This problem assesses students' ability to identify and describe patterns of increase and/or decrease from a table or graph and their ability to make connections between situation, graph, and table. This problem also assesses their ability to reason about growth situations in terms of slope, maximum and minimum, range, decrease, and increase.

Students may find it helpful to copy and cut out each of the notes, reorganize them, and paste them on a sheet of paper (see the solutions column).

If you notice that some students misinterpret tables as ratio tables, you might discuss how to tell whether or not a table is a ratio table (see About the Mathematics).

Growth Charts

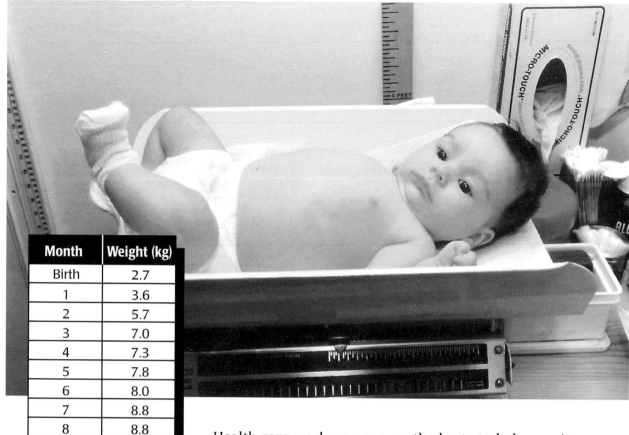

Month	Weight (kg)
Birth	2.7
1	3.6
2	5.7
3	7.0
4	7.3
5	7.8
6	8.0
7	8.8
8	8.8
9	8.8
10	9.3
11	9.6
12	10.5
13	10.3
14	11.3
15	12.0
16	12.4
17	12.9
18	13.1
19	12.9
20	10.5
21	9.2
22	9.5
23	12.0
24	13.0
25	13.6
26	13.5
27	14.0
28	14.2

Health care workers use growth charts to help monitor children's growth.

30. Why is it important to monitor a child's growth?

The chart on the left shows the weight records, in kilograms (kg), of a 28-month-old boy.

31. How would you describe this boy's growth? Do you think that he received adequate nutrition?

The growth charts that follow show normal ranges for the weights and lengths of young children. The normal growth range is indicated on the charts by curved lines. A child whose weight is not within the range may be suffering from malnutrition or some other problem.

30. Answers will vary. Sample student responses:

- To make sure the child is healthy.

- To make sure the child does not have a disease.

- If the child weighs too little, it is not healthy.

31. Answers will vary. Sample response:

He seemed to be gaining weight and growing until he was 18 months old. From 18 to 22 months old, he lost weight. Then from 22 months old to 28 months old, he started gaining weight again.

Overview Students investigate a table showing weight records for a young child from birth up to the age of 28 months.

About the Mathematics All growth situations in the preceding problems showed increases, whether in height, size, or some other measurement. On this page, decrease is introduced in the form of weight loss. Using the records in the table, one can distinguish increase from decrease by using positive and negative numbers for these differences:

Month	Weight (in kg)	
11	9.6	⟩ + 0.9
12	10.5	⟩ − 0.2
13	10.3	⟩ + 1.0
14	11.3	

Planning Students may work on problems **30–31** in pairs or in small groups. When they have finished, you might discuss these problems in class.

Comments about the Problems

30. You might have students discuss which answers are reasonable.

WEIGHT GROWTH CHART FOR BOYS
Age: Birth to 36 months

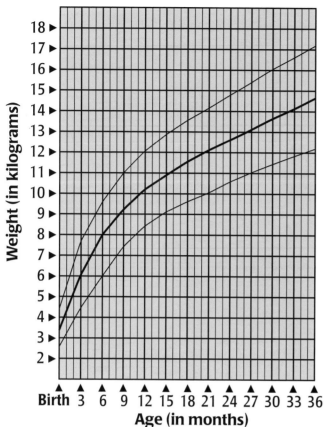

Weight (in kilograms)

18 ▶
17 ▶
16 ▶
15 ▶
14 ▶
13 ▶
12 ▶
11 ▶
10 ▶
9 ▶
8 ▶
7 ▶
6 ▶
5 ▶
4 ▶
3 ▶
2 ▶

Birth 3 6 9 12 15 18 21 24 27 30 33 36

Age (in months)

© Am. J. Clin. Nutr.
American Society for Clinical Nutrition

32. What does the chart on the left measure? What does the chart on the bottom of the page measure?

In both charts, there is one curved line that is thicker than the others.

33. What do these thicker curves indicate?

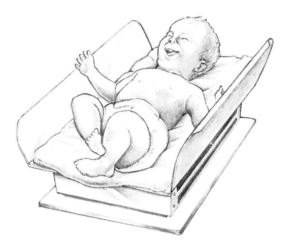

LENGTH GROWTH CHART FOR BOYS
Age: Birth to 36 months

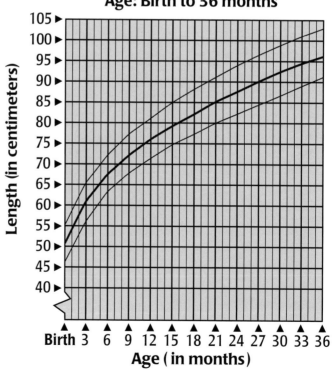

Length (in centimeters)

105 ▶
100 ▶
95 ▶
90 ▶
85 ▶
80 ▶
75 ▶
70 ▶
65 ▶
60 ▶
55 ▶
50 ▶
45 ▶
40 ▶

Birth 3 6 9 12 15 18 21 24 27 30 33 36

Age (in months)

© Am. J. Clin. Nutr.
American Society for Clinical Nutrition

32. The chart on the left measures the weight in kilograms of boys from birth up to the age of 36 months, or three years. The chart at the bottom of the page measures the length in centimeters over the same period.

33. The thicker curves show the median weights and lengths for boys. Half, or 50%, of the boys are heavier or taller than the thick line, and half, or 50%, are lighter or shorter.

Materials transparencies of the charts on Student Book page 15, optional (one per class); overhead projector, optional (one per class)

Overview Students investigate a weight growth chart and a length growth chart.

About the Mathematics The curves in the charts show the results of measurements of the weights and lengths of a group of young boys. The dark line in the center of each set of lines is the median weight or height for boys, or the 50th percentile. The bands show the distribution of the middle 90% of children (from 5% to 95%). Students may remember the concept of median from the unit *Dealing with Data.*

Planning You may want to make transparencies of the charts and use them to lead a class discussion. Students may work on problems **32** and **33** as a whole-class activity.

Comments about the Problems

32. Students should note that this chart shows a record of growth for boys. A different chart exists for girls. You might want to discuss why there would be different charts for boys and girls.

34. Graph the weight records that you used in problem **31** on the weight growth chart on **Student Activity Sheet 4.**

35. Study the graph that you made. What conclusions can you draw from it?

A baby boy's length at birth was 48 centimeters. At 28 months, his length was 89 centimeters.

36. Use the length growth chart on **Student Activity Sheet 4** to record what you think his length might have been for each of the months between birth and 28 months. Explain how you decided where to put the dots.

37. What records are more important to keep: height or weight? Why do you think so?

Below are four weekly weight records for two children. The records begin when the children are one year old.

	Week 1	Week 2	Week 3	Week 4
Samantha's Weight (in kg)	11.8	11.6	11.3	10.9
Hillary's Weight (in kg)	10.5	10.0	9.7	9.6

38. Although both children are losing weight, which one would you worry about more? Why?

34.

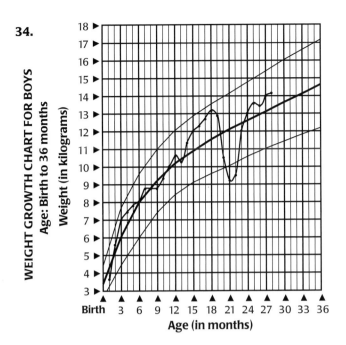

WEIGHT GROWTH CHART FOR BOYS
Age: Birth to 36 months
Weight (in kilograms)
Age (in months)
Birth 3 6 9 12 15 18 21 24 27 30 33 36

35. Answers will vary. See Hints and Comments for sample response.

36. Sample chart:

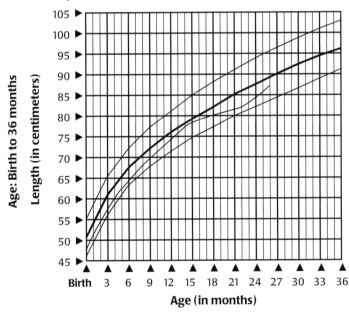

Age: Birth to 36 months
Length (in centimeters)
Age (in months)
Birth 3 6 9 12 15 18 21 24 27 30 33 36

Explanations will vary. Sample student response:

My graph stayed a bit at the lower part of the band because at birth the baby didn't weigh so much, so I thought it was a small baby.

37. Answers and explanations will vary.

38. Answers will vary. See the Hints and Comments column for sample responses.

Materials Student Activity Sheet 4 (one per student)

Overview Students graph weight records on a weight growth chart. They also create a graph to show an increase in length. Then they investigate the weight records of two children in a table.

Planning Students may work on problems **34–37** in pairs or in small groups. You might have a class discussion about their solutions to problems **35–37** before they continue with problem **38**. Students may work individually on problem **38**. This problem may also be used as informal assessment.

Comments about the Problems

34. You might point out that most babies are not weighed every month.

35. Sample student response:

The baby's weight gain was normal until he was 18 months old. At this point he lost weight very fast. Three months later, he gained the weight back almost as fast. After that point he seemed to gain weight normally.

36. Some students may use some of the records in the table of Dean's height on Student Book page 9 as reference points for height at birth and at ages 1, 2, and 3. Be sure students see that as far as children are concerned, length can only increase or stay the same, whereas weight can increase and decrease.

37. Some students may say weight is more important, since height doesn't decrease and therefore won't indicate a problem.

38. Informal Assessment This problem assesses students' ability to identify and describe patterns of increase and/or decrease from a table or graph and to reason about situations of growth in terms of slope, maximum and minimum, range, decrease, and increase.

You might point out that the seriousness of the weight loss might depend on the child's height.

Samantha's weight loss could be considered more serious because it is more each week than the week before. Hillary's weight loss could be more serious because she weighs less than Samantha.

WATER FOR THE DESERT

In many parts of the world, you can find deserts by the sea. Since there is a water shortage in the desert, you might think that you could use the nearby sea as a water source. Unfortunately, seawater contains salt that would kill the desert plants.

Some scientists are investigating ways to bring ice from the Antarctic to the desert. The ice from an iceberg is made from fresh water. It is well packed and can be easily pulled by boat. However, there is one problem: the ice would melt during the trip, and the water from the melted ice would be lost.

There are different opinions about how the iceberg might melt during a trip. The graphs on page 18 illustrate three of these opinions.

Overview Students are introduced to problems in a new context, icebergs as a source of fresh water. There are no problems on this page for students to solve.

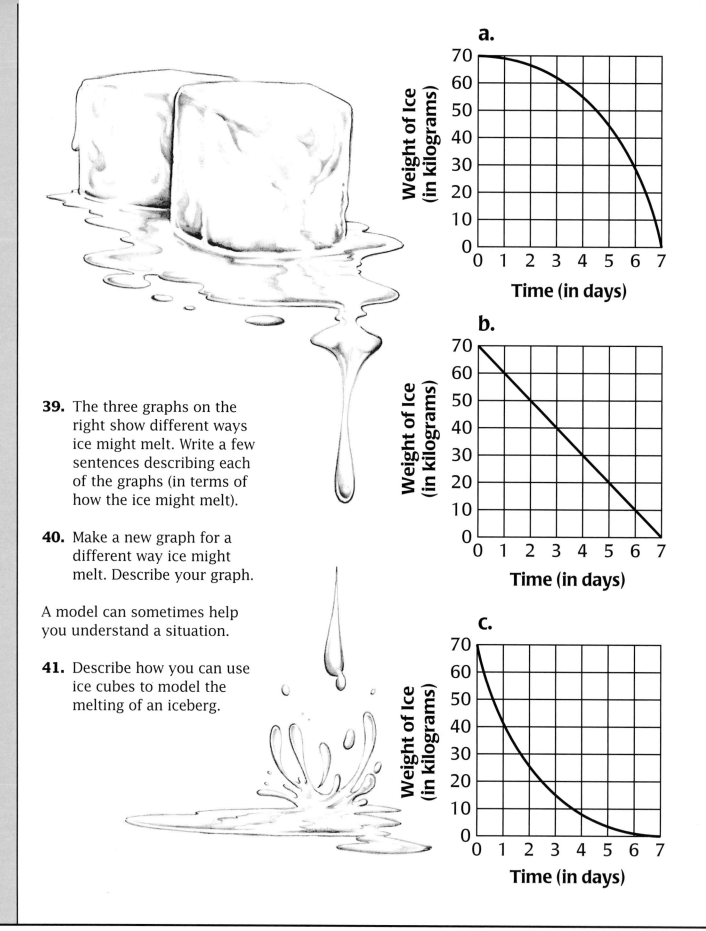

a.

b.

c.

39. The three graphs on the right show different ways ice might melt. Write a few sentences describing each of the graphs (in terms of how the ice might melt).

40. Make a new graph for a different way ice might melt. Describe your graph.

A model can sometimes help you understand a situation.

41. Describe how you can use ice cubes to model the melting of an iceberg.

39. Descriptions will vary. Sample responses:

Graph "a" shows a situation in which every day there will be more of the volume melting than the day before. Or, the ice is melting quicker and quicker.

In graph "b," the volume decreases by the same amount every day.

Graph "c" shows that every day there is less ice melting than the day before. Or, the ice is melting slower and slower.

40. Graphs will vary. Sample graphs:

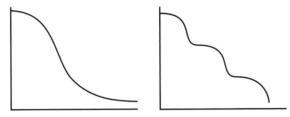

Explanations will vary. Some students might say that the latter graph reflects the temperature changes during the day and night.

41. Answers will vary. Sample response:

To model the melting of an ice cube, you might freeze a toothpick into an ice cube and then put it into a measuring cup with water. The change in water level every minute (after removing the ice cube) could then be measured. In this way you get the volume of the melting water, which does not have exactly the same volume as when it was frozen.

Overview Students explain graphs that show different ways an iceberg might melt. They create another graph. Then they describe how ice cubes can be used to model a melting iceberg.

About the Mathematics The rate at which an iceberg melts may depend on different circumstances: the temperature of air and water, the wind, the movement of the water, and the shape and size of the iceberg. The shape of the iceberg is quite important. A flat iceberg would melt more quickly than an iceberg with the same volume but in the shape of a sphere (or a cube), because its surface area is greater. This can be shown by comparing a block and a cube with the same volume as shown below:

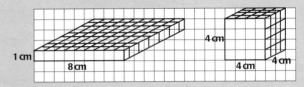

volume = 64 cm^3 volume = 64 cm^3
surface area = 160 cm^2 surface area = 96 cm^2

Planning Students may work individually on problems **39–41.** You may want to assign problems **39–41** as homework. If time is a concern, you can omit problems **40** and **41.** Discuss students' answers to problem **39** in class. It may be very instructive to the students when they hear from each other how the different graphs can be described.

Comments about the Problems

39–41. **Homework** These problems can be assigned as homework.

39. If students are having difficulties, then you might pay extra attention when discussing the Summary, and take more time to look at different possible patterns of decrease.

Summary

By looking at a graph, you can see whether something is increasing or decreasing over time.

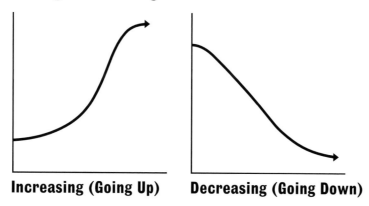

Increasing (Going Up)   **Decreasing (Going Down)**

If you have a table of numbers, you can see whether the numbers are getting bigger (increasing) or getting smaller (decreasing) over time. You can also draw a graph using the information in a table.

The shape of a graph shows how a value increases or decreases. A table can show this too, but you have to do some calculations first.

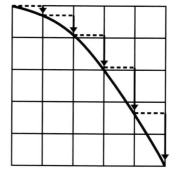

The graph on the right and the table below show a value that is decreasing more and more.

Days	0	1	2	3
Weight	25	23	19	10

−2   −4   −9

Summary Questions

42. Using a graph and a table, show how a value can increase more and more.

42. Graphs and tables will vary. Sample graph and table:

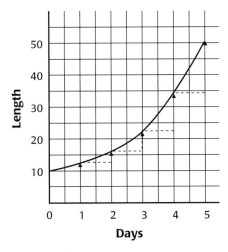

Days	0	1	2	3	4	5
Length	10	12	17	23	35	50

+2 +5 +6 +12 +15

Overview In the Summary, students see how the shape of a graph and data in a table can show that a value decreases more and more rapidly over time. Then they give an example of a graph and a table to show a value that increases more and more rapidly over time.

About the Mathematics When the pattern of increase or decrease is investigated, it is important that the differences be calculated for equal time intervals, and that the same time intervals are used in corresponding tables and graphs.

The dotted lines in the graph on Student Book page 19 show these time intervals, and the arrows show the differences. (See the graph in the solutions column.)

Another way to display the differences is by the following diagram:

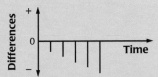

The lines in the diagram are related to the arrows on the graph in the Summary, and they point downward to indicate decreasing values. Although this type of diagram was useful at the beginning of this section because it represented real-world situations, using it here is probably too abstract for the students to understand.

Planning Read and discuss the Summary with the whole class. As extension questions for your discussion, you might say: *Tracy looked at the first graph on the left and said "It's increasing at first, but decreasing towards the end."* Ask students whether they agree or disagree with Tracy and explain why. [The graph is still increasing toward the end, but it is increasing more slowly.] You might ask a similar question for the graph on the right.

Students may work on problem **42** in pairs or in small groups. Discuss students' responses before students continue with the problems on the next page.

Comments about the Problems

42. You may want to encourage students to sketch the graph first and then refine it by making a table. The table and the graph do not necessarily have to match as long as each shows a quantity that is increasing more and more rapidly. If students find this difficult, you may suggest that they make a table first and then draw a graph using the table.

Summary Questions, continued

In newspapers, you often find headlines like these.

On the right are four different graphs.

43. Which graph fits the headline on the left?

44. Which graph fits the headline on the right?

45. Make up headlines for the other two graphs.

A

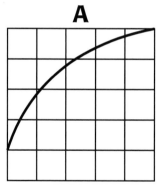

B

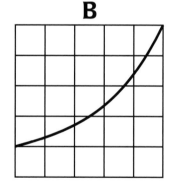

C

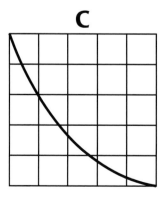

D

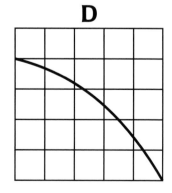

43. Graph B: The headline "Population of Seattle Growing Faster and Faster" can be shown in the graph:

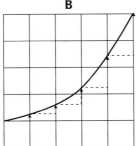

B

44. Graph D: The headline "Price of Calculators Dropping Faster and Faster" can be shown in the graph:

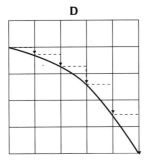

D

45. Answers will vary. Sample responses:

Graph A: "Rise in Crime Slows Down"

Graph C: "Decline in Joblessness Slows Down"

Overview Students find out which of the four graphs match each of the two given headlines. They then make up headlines for the two remaining graphs.

Planning Students may work individually on problems **43–45.** Problems **43–45** may also be used as informal assessment. You may want to have a short class discussion about students' solutions. After students complete Section A, you may assign appropriate activities in the Try This! section, located on pages 54–57 of the Student Book, for homework.

Comments about the Problems

43–45. Informal Assessment These problems can be used to assess students' ability to identify and describe patterns of increase and/or decrease from a graph and to make connections between situation, graph, and table.

43. You might ask students to discuss their strategies. Some students may first decide that it is a situation of increase, (narrowing it down to either graph A or graph B), and then analyze the pattern of increase in each graph and draw their conclusions.

44. Be sure that students use correct descriptions for the patterns of increase and decrease. You might ask five students to write their headlines for graph A on the board and then discuss these headlines. Then you might ask whether anyone in the class has described the pattern of increase in different words.

Work Students Do

In this section, students apply the concept of periodic, or cyclical, graphs to a variety of real-world situations. Changes in water level during high and low tides are explored in the context of a fishing trip, a nature walk in The Netherlands, and the area near the Golden Gate Bridge in California. Students plot changes in water level over time on graphs and compare the resulting periods and cycles. Next, students continue to investigate how periodic graphs are used to represent repeating patterns in the context of air temperature changes in a room, body temperature changes in a camel, blood pressure changes in a person, and speed changes in cars running laps around a racetrack.

Goals

Students will:

- identify characteristics of periodic graphs;
- make connections between situation, graph, and table;
- use algebraic models to represent realistic situations;*
- recognize the power of graphs and/or tables for representing and solving problems;*
- identify and describe patterns of increase and/or decrease from a table or graph;*
- reason about situations of growth in terms of slope, maximum and minimum, range, decrease, and increase.*

  * *These goals are addressed in this section and assessed in other sections in the unit.*

Pacing

- approximately three 45-minute class sessions

Vocabulary

- cycle
- period
- periodic

About the Mathematics

Periodic or cyclical graphs are informally introduced in this section. The period of a periodic function is the time or distance it takes for the graph to repeat, and the cycle is a part of the graph that repeats. A cycle can be chosen at different places on a graph, as long as a cycle covers a period.

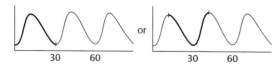

Periodic functions occur frequently in nature, including in the human body (for example, the volume of the lungs or blood pressure over time). Students use tables, verbal descriptions, and graphs to represent these repeating patterns.

Materials

- Student Activity Sheets 5 and 6, pages 147 and 148 of the Teacher Guide (one of each per student)
- graph paper, pages 49, 53, 61, and 63 of the Teacher Guide (four sheets per student)
- transparency of the tidal graphs on Student Book page 22, page 50 of the Teacher Guide, optional (one per class)
- overhead projector, page 51 of the Teacher Guide, optional (one per class)

Planning Instruction

You might have a brief class discussion to introduce tides, as some students may not be familiar with tides. Question them about their knowledge of tides or discuss how tides work. Some students may remember the use of tidal graphs to solve problems about ships entering a harbor from the unit *Tracking Graphs.*

Have students work on problem 4 as a whole class and work individually on problems 11–17. Problems 8–10 may be done by students individually or working in pairs. They may work on the remaining problems in pairs or small groups.

Problems 18–21 are optional. If time is a concern, you may omit these problems or assign them as homework.

Homework

Problems 5–7 (page 52 of the Teacher Guide) and problems 14–17 (page 58 of the Teacher Guide) can be assigned as homework. The Extensions (pages 53, 59, and 63 of the Teacher Guide) may also be assigned as homework. After students finish Section B, you may assign appropriate activities from the Try This! section, located on pages 54–57 of the *Ups and Downs* Student Book. The Try This! activities reinforce the key mathematical concepts introduced in this section.

Planning Assessment

- Problem 11 may be used to informally assess students' ability to make connections between situation, graph, and table, and to identify characteristics of periodic graphs.
- Problems 12, 13, and 21 may be used to informally assess students' ability to identify characteristics of periodic graphs.

Fishing

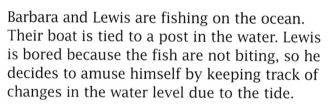

Barbara and Lewis are fishing on the ocean. Their boat is tied to a post in the water. Lewis is bored because the fish are not biting, so he decides to amuse himself by keeping track of changes in the water level due to the tide.

He makes a mark on the post every 15 minutes. He made the first mark (at the top) at 9:00 A.M.

1. What do the marks tell you about the way the water level is changing?

2. Make a graph that shows how the water level changed during this time?

3. What can you say about how the graph will continue? (Think about the tides of the ocean.) You don't need to graph it.

1. Answers will vary. Sample responses:

   The water level is going down more and more.

   The water level is decreasing at a faster rate.

2.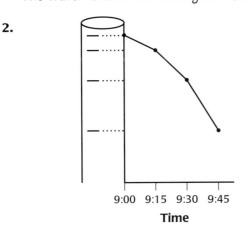

   9:00 9:15 9:30 9:45
   Time

3. Answers will vary. Students' responses should indicate that the graph may continue downward for a number of hours and then eventually begin to rise. Some students may say that the graph will repeat the same pattern over time.

Materials graph paper, optional (one sheet per student)

Overview Students use marks on a post to draw a graph that shows how the water level changes over time.

About the Mathematics The marks on the post showing the differences in water height are similar to the marks on the door and the core samples of Section A. They all show a pattern of change over time. Note that the marks on the post here indicate a change in the rate of decrease.

This section reinforces the main concepts studied in the previous section. The context of this section—the change of water level, which is influenced by the tides—also produces continuous graphs. Because tide phenomena are periodic events, the graphs also show a regular, repetitive pattern.

Planning You may wish to start this section with a short discussion about tides. Students may work on problems **1–3** in pairs or in small groups. Be sure to discuss students' solutions and strategies for these problems.

Comments about the Problems

1. Make sure that students realize that the marks on the post were made consecutively from top to bottom.

2. Some students may trace the marks from the post (from Student Book page 21) on the edge of a strip of paper and use this strip to draw the four points of the graph. The points of the graph can then be connected. See the solutions column. Some students may want to use graph paper to draw the graph.

3. Some students may remember the tidal graphs from the unit *Tracking Graphs*. Some students might not be familiar with tides. You might discuss how the water level goes up and down.

   Note: The graph on Student Book page 22 shows that the water level could fall by as much as 1.6 meters over 5 hours!

HIGH TIDE, LOW TIDE

During low tide in The Netherlands, naturalists lead hikes along the Dutch Shallows (de Wadden), or coastal tide flats. They point out the various plant and animal species that live in this unique environment. Seaweed, oystercatchers, curlews, plovers, mussels, jellyfish, and seals are some of the plants and animals hikers see on a walk.

Walkers do not always stay dry during the walk; sometimes the water may even be waist-high. Walkers carry dry clothing in their backpacks and wear tennis shoes to protect their feet from shells and sharp stones.

On the right are graphs of the tides for the Dutch Shallows for two days in April. Walking guides recommend that no one walk in water deeper than 45 centimeters above the lowest level of the tide.

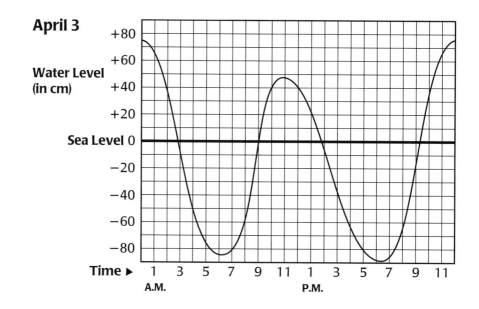

4. When is it safe to walk on the Dutch Shallows on these two days in April?

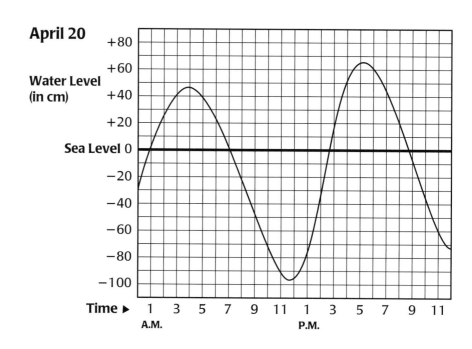

4. The safe times to walk on the Dutch Shallows are:

April 3: from about 3:45 A.M. until about 8:20 A.M. (about 4.5 hours) and from about 3:15 P.M. until about 8:30 P.M. (about 5.25 hours)

April 20: from about 9:15 A.M. until about 1:45 P.M. (about 4.5 hours)

Note: The levels of low and high tides are different at different times.

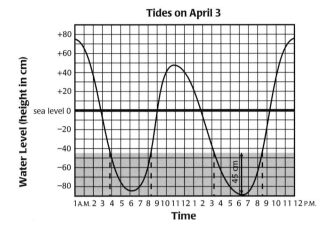

Tides on April 3

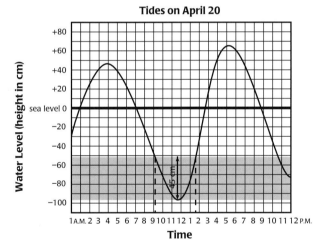

Tides on April 20

Materials transparency of the tidal graphs on Student Book page 22, optional (one per class); overhead projector, optional (one per class)

Overview Students interpret a tidal graph to solve a problem about walking along the Dutch Shallows.

About the Mathematics Students may have seen a problem similar to this one in *Tracking Graphs*, involving tides and the water depth required for a ship to enter a harbor. By drawing a horizontal line in the tidal graph, the times when the ship could enter the harbor could easily be found. The same strategy can be used here. Note, however, that there is a difference between these graphs and those in *Tracking Graphs*. Here the water level is related to the mean; the zero is at sea level, whereas in *Tracking Graphs*, the water level is measured from the bottom of the sea. This difference has no bearing on the shape of the graph.

Planning You might read this page and complete problem **4** together as a whole-class activity. Ask students whether or not these tidal graphs tell anything about the depths of the water. You might want to make a transparency of the tidal graphs shown on Student Book page 22 and have some students show how they used the graphs to solve problem **4** on the overhead projector.

Comments about the Problems

4. If students have difficulty starting this problem, you might ask the following questions:

- *Where on the graph is the lowest level of the tide?* [6:30 P.M. for April 3 and 11:30 A.M. for April 20]

- *Where is the water 45 centimeters above the lowest level?* [all points at −43 cm for April 3 and at −50 cm for April 20]

Then have students read the last sentence of the text again. Ask them what they need to find out now and how they can use the above information.

The solution can be found by shading the section of the graph (or by drawing a horizontal line at a height) that is 45 centimeters above the lowest level of the tide. See the solutions column.

Note: It is only safe to walk along the Dutch Shallows after sunrise; this area could be dangerous in the dark.

Golden Gate Bridge

In different parts of the world, the levels of high and low tide vary. The amount of time between the tides may also vary. On the right is a tide schedule for the area near the Golden Gate Bridge in San Francisco, California.

5. Use the information in this table to sketch a graph of the water levels near the Golden Gate Bridge for these three days. Use **Student Activity Sheet 5** for your graph.

6. Describe how the water level changed.

7. Compare your graph with the two graphs on page 22. What similarities and differences do you notice?

High and Low Tides at Golden Gate Bridge

Date	Low	High
Aug. 7	2:00 A.M./12 cm 1:24 P.M./94 cm	9:20 A.M./131 cm 7:47 P.M./189 cm
Aug. 8	2:59 A.M./6 cm 2:33 P.M./94 cm	10:18 A.M./137 cm 8:42 P.M./189 cm
Aug. 9	3:49 A.M./0 cm 3:29 P.M./91 cm	11:04 A.M./143 cm 9:34 P.M./186 cm

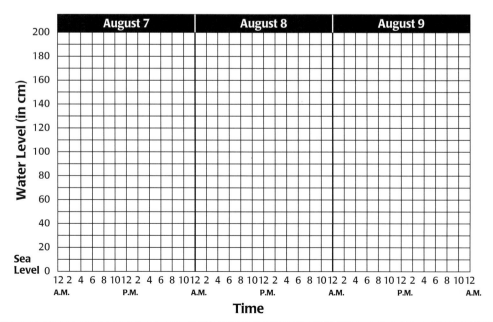

5. Sample graph:

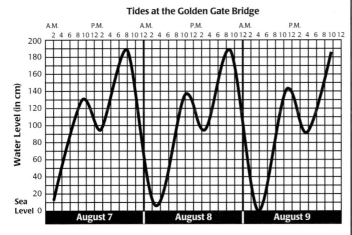

Tides at the Golden Gate Bridge

6. Answers will vary, and some descriptions may be more detailed than others. For example, students may include extensive information about times and heights. Sample responses:

On the first day, the tide rose for about seven hours and at 9:20 A.M. it was high tide. Then it fell for about four hours. Then the tide rose again for about six hours. When high tide was reached at about 8:00 P.M., the level was higher than the high tide level in the morning. Then for the rest of the day the tide fell. On the second and third days, the changes in the water level show the same pattern as on the first day.

There are two high and two low tides each day. It seems like the low tides are getting lower and the high tides are getting higher.

7. Answers will vary. Sample responses:

All the graphs show that there are two high tides and two low tides for each day. The difference is that in San Francisco, the water levels for the two high tides in a single day differ quite a bit. The same thing is true for water levels for the low tides. In the Dutch Shallows in The Netherlands, the water levels for the high tides on a single day are almost the same.

The graph of the tides in San Francisco varies more than the graph of the tides in the Dutch Shallows in The Netherlands.

The water level in The Netherlands changes from below sea level to above sea level; the water level in San Francisco is always at or above sea level.

Materials Student Activity Sheet 5 (one per student); graph paper (one sheet per student)

Overview Students use the data in a table about the times and levels of low tide and high tide to sketch a graph of the water level at the Golden Gate Bridge over a period of three days. Then they compare this graph with those on the previous page.

Planning Students may work in pairs on problems **5–7.** You may want students to finish these problems as homework.

Comments about the Problems

5–7. Homework These problems may be assigned as homework.

5. Students use the eight given points to draw a line graph. The graph should not have sharp zigzag edges, since the water level increases and decreases gradually. Students' graphs will vary in precision and neatness.

It is important to discuss whether or not this graph could have a vertical line for one section. Ask students what it would mean to have a vertical line in the graph. (This would mean that the water level dropped or rose a certain number of centimeters in zero seconds.) Observe that no tidal graph shows curves like this:

(These graphs show three different water levels at the same time, which is impossible.)

To help students get started, discuss how to plot points for the first two hours of the day on August 7.

6. Students' answers should include a description of the two (relative) high tides and low tides on one day, and the repeating pattern.

7. Differences in water levels of high tides and low tides at the Golden Gate Bridge and the Dutch Shallows are due to the differences in the shapes of the coastlines near the two bodies of water.

Extension If you live near bodies of water that have tidal fluctuations, you can obtain local times for high and low tides. Students could then use this information to draw a tidal graph.

The Air Conditioner

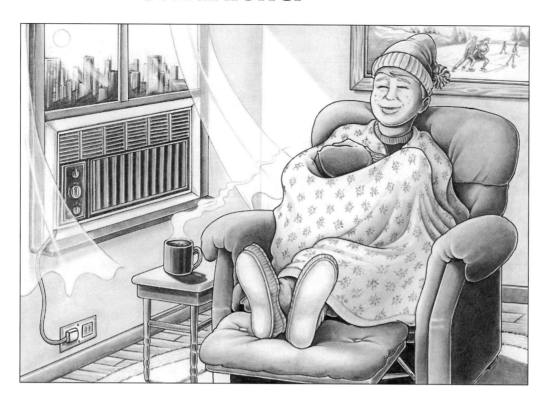

Suppose the graph below shows the temperature changes in an air-conditioned room.

8. Describe what is happening in the graph. Why do you think this is happening?

9. What numbers might you write along the horizontal and vertical axes? Label the axes on **Student Activity Sheet 6** with these numbers.

Vertical axis: Temperature

Horizontal axis: Time (in minutes)

The graphs you have seen in this section have one thing in common: they have a shape that repeats. A repeating graph is called a *periodic* graph. The amount of time it takes for a periodic graph to repeat is called a *period* of the graph. The portion of the graph that repeats is called a *cycle.*

10. a. How long (in minutes) is a period in the above graph?

   b. On **Student Activity Sheet 6,** color one cycle on the graph.

8. Answers will vary. Some students may comment that the temperature rises when the air conditioner is off and falls when it is on.

9. Answers will vary. Sample response:

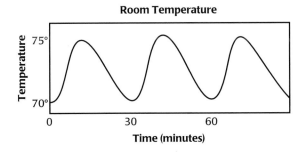

Room Temperature

10. a. Answers will vary, depending on the numbers students choose to use along the horizontal axis. A period in the graph shown above is 30 minutes.

b. Answers will vary, depending on where a student starts to color a cycle. A cycle is correctly shown if it covers the time period that corresponds to students' answers for problem **10a.**

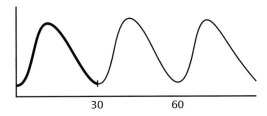

or

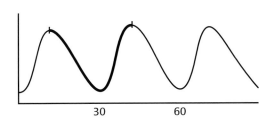

Materials Student Activity Sheet 6 (one per student)

Overview Students interpret a graph that shows a repeating pattern of temperature changes in an air-conditioned room.

About the Mathematics The concepts of a *periodic* graph, a *period*, and a *cycle* are made explicit on Student Book page 24. Although these are new mathematical concepts, students may already have some understanding of periodic events, since repeating patterns are frequently found in science and nature. For example, sound and light waves show repeating patterns that can be illustrated using a periodic graph. High and low tides, phases of the moon, as well as the human heartbeat are also characterized by repeating patterns that make periodic graphs. Note that sometimes a phenomenon approximates a regular pattern but is not exactly periodic. In such cases, the periodic function serves as a model.

Planning You might want students to work in pairs or individually on problems **8–10** to see how well each student is able to recognize the repeating pattern. Discuss students' answers to these problems.

Comments about the Problems

8. By this point in the unit, students should be able to think about periodicity in terms of both the real-world context and the graph.

9. Discuss the factors that might affect which numbers are written along each axis. Some students may suggest that the time intervals for each period would be smaller on a really hot day or if a person wanted the room to be cold.

10. This is the first problem in which students must identify the period and cycle in a periodic graph. If students are having difficulty, you might draw a different example of a periodic graph and discuss these concepts with students. Point out that a graph's cycle can be colored at different places on the graph curve. Once one cycle is colored, it makes it easy to identify the period of the graph: the time interval along the horizontal axis that it takes to complete one cycle.

The Camel

The body temperature of a camel changes greatly over time. As the desert air warms during the day, the camel's body temperature increases. The camel starts sweating when its body temperature reaches 40°C. During the night, the desert air cools off, and the camel's body temperature decreases. The lowest temperature will be reached at about 4:00 A.M.

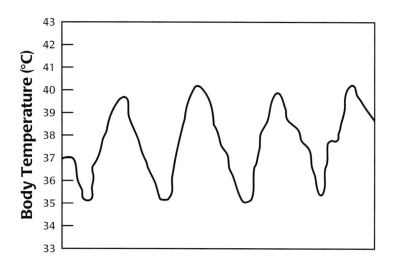

Use the graph on **Student Activity Sheet 6** to complete the problems below.

11. Label the horizontal axis.

12. How long is one period of the graph?

13. Color one cycle on the graph. Describe what is happening to the camel's body temperature during this period.

11. Students' graphs should label the lowest point as 4:00 A.M. and the highest as about 4:00 P.M. Students may use different time intervals (count by twos, fours, and so on) on the horizontal axis.

12. One period is 24 hours, or one day.

13. Answers will vary, depending on the particular cycle that is chosen.

Sample response:

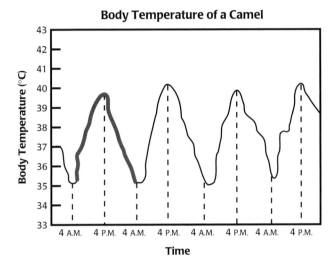

Body Temperature of a Camel

The body temperature of the camel gradually rises from 4:00 A.M. until 4:00 P.M. At this point, it begins to fall steadily until 4:00 A.M.

Materials Student Activity Sheet 6 (one per student)

Overview Students identify one period and cycle on a periodic temperature graph.

Planning You may want to have students work individually on problems **11–13** to see how well each student understands and uses the concepts that were explained on the previous page. Problems **11–13** may be used as informal assessment. Discuss students' solutions to problems **12** and **13** in the following class session.

Comments about the Problems

11. Informal Assessment This problem assesses students' ability to make connections between situation, graph, and table and to identify characteristics of periodic graphs.

12–13. Informal Assessment These problems assess students' ability to identify characteristics of periodic graphs. This graph, taken as a whole, shows a periodic phenomenon. Some students may notice, however, that not all the "ups and downs" in the graph are similar. You may want to point out that small fluctuations may be due to weather changes, how often and how much water the camel drinks, or other factors.

Blood Pressure

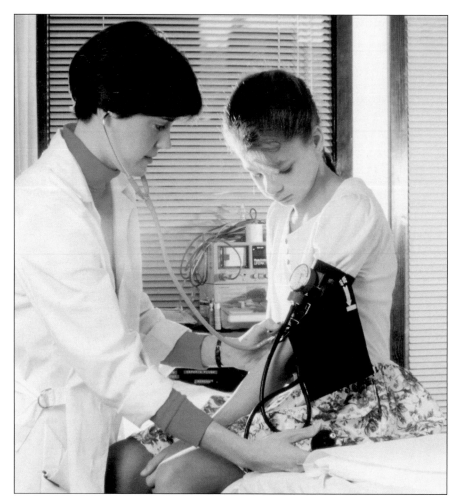

Your heart pumps blood throughout your system of arteries. When doctors measure blood pressure, they usually measure the pressure of the blood in the artery of the upper arm.

Your blood pressure is not constant. The graph below shows how blood pressure may change over time.

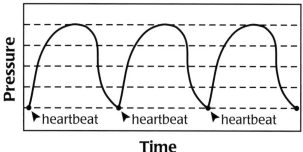

Time

14. What can you tell about blood pressure just before a heartbeat?

15. What happens to blood pressure after a heartbeat?

16. Is this graph periodic?

17. Suppose this is a graph of your own heartbeat. How many seconds long is a period?

14. Answers will vary. Sample response:

The blood pressure is at its lowest point just before the heart beats.

15. Answers will vary. Sample response:

After a heartbeat, the blood pressure rises rapidly, then rises more slowly, and finally stops rising. The pressure then begins to fall, slowly at first, then more rapidly, then more slowly again until it reaches its lowest value.

16. Yes, this is a periodic graph because the cycle repeats.

17. Answers will vary and will depend on how long it takes for the student's heart to beat one time. The answer would be one second if the student's heart beats 60 times per minute.

Overview Students investigate a graph that shows how blood pressure can change over time. They also determine the period of their own heartbeats.

Planning Before students begin problem **17,** make sure they know how to take their pulse. These problems may be assigned as homework. You may want students to work individually on problems **14–17.**

Comments about the Problems

14–17. Homework These problems may be assigned as homework.

16. You might discuss how the graph would change if the person had been running and the heartbeat were twice as fast.

17. Students must count the number of times their hearts beat in a specific time period in order to calculate the period. They may count the number of beats in a 10-, 20-, or 30-second time period or for one full minute.

Some students may find it hard to calculate a period, especially since their hearts probably don't beat exactly 60 times per minute. A ratio table may be helpful. For example:

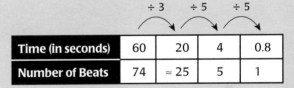

	÷ 3	÷ 5	÷ 5
Time (in seconds)	60	20	4	0.8
Number of Beats	74	≈ 25	5	1

In this example, a period is 60/74 or about 0.8 seconds.

Extension You may have students work in pairs or small groups on the following activity:

1. Use a reference book or experiment to find the following information:

- How many times do you inhale and exhale in one minute?

- How much air do you inhale when you breathe normally?

- What is the volume of your lungs?

2. Make a graph that shows the volume of your lungs over a time period of 15 seconds.

3. If your graph is periodic, how long is a period? If it is not periodic, explain why.

The Racetrack

A group of race-car drivers is at a track getting ready to practice for a big race. Below you see a picture of the racetrack.

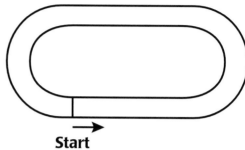

Start

18. Sketch the track. Color it to show where the drivers should speed up and where they should slow down. Include a key to show the meanings of the colors or patterns you chose.

19. Make a graph of the speed of a car during three laps around the track. Label your graph like the one on the right.

20. Explain why your graph is or is not periodic.

Speed

Distance along the Track

18. Drawings will vary. Sample drawing:

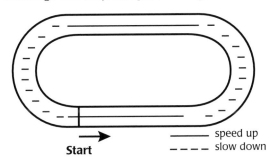

speed up
slow down

Start

19. Answers will vary, but the graph should correspond with the colored parts of the track in problem **18.** Sample response:

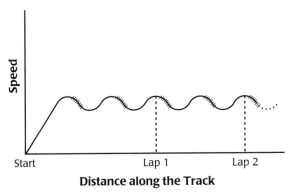

Distance along the Track

Note: Graph shows only 2 laps.

20. Explanations will vary. Sample explanation:

My graph is periodic. It shows a repeating pattern between the beginning and the end of the race.

Materials graph paper (one sheet per student)

Overview Students sketch a racetrack to indicate where race car drivers should speed up and slow down. They also make a graph that shows the speed of a race car during three laps around the track.

About the Mathematics The periodic graph on this page shows the speed of a car over time. Students might remember making a similar graph in the unit *Tracking Graphs*.

Planning You may want to begin with a short discussion about the context, racetracks, and talk about graphs that show speed over time. If students have difficulty distinguishing between graphs that show distance over time and speed over time, you might want students to do an activity similar to the one they did in the unit *Tracking Graphs*. In that unit, they created a speed graph for a car ride with miles per hour on the vertical axis and times on the horizontal axis.

Students may work in pairs on problems **18–20.** These are optional problems and may be omitted or assigned as homework if time is a concern.

Comments about the Problems

18. The purpose of this problem is to get students to begin thinking about where a race car driver will speed up or slow down before students actually make a graph showing the speed over time. In general, race car drivers speed up halfway through the curved areas and on the straightaways. Race cars may be traveling at a constant speed at the end of the straightaways if the cars are being driven at their maximum speed.

19. Students may use the racetrack sketch they drew in the previous problem to help them make their speed graphs here.

20. Some students may not recognize that the speed graph is periodic, since the graph would not continue to repeat forever. You may need to help students identify what one period is in this graph: the distance it takes to complete one cycle on the graph.

Summary

A periodic graph shows a repeating pattern.

A period on the graph is the length of time or the distance required to complete one *cycle*, or the part of the pattern that is repeated.

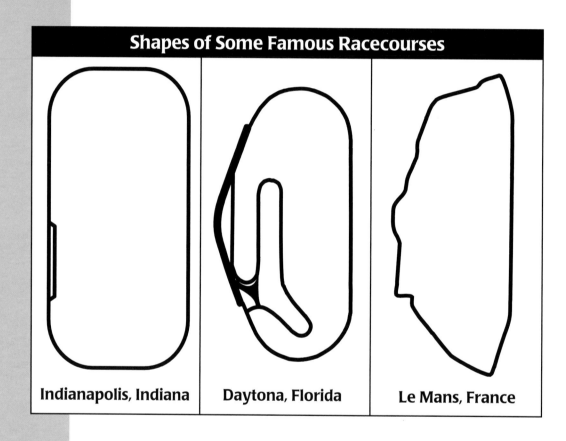

Shapes of Some Famous Racecourses

Indianapolis, Indiana | **Daytona, Florida** | **Le Mans, France**

Summary Questions

21. Design a racetrack of your own. On another piece of paper, draw a graph to show how the speed of a car will vary as it moves around one lap of your track. Pass your graph to your neighbor and have him or her reconstruct the shape of your racetrack.

21. Drawings and graphs will vary. Sample drawings and graphs:

Racetrack

Speed

Distance along the Track

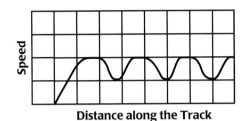

or

Racetracks

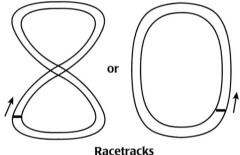

Speed

Distance along the Track

Materials graph paper (one sheet per student)

Overview Students read and discuss the Summary, which reviews the main concepts of this section. Then they design a racetrack and draw a possible speed graph for a car on this track.

Planning You might discuss the Summary with the whole class as follows. Ask one student to draw an example of a periodic graph on the board. Then ask another student to explain why the graph is periodic. If the axes aren't labeled yet, you may ask a third student to label them. Another student can be asked to color a cycle, and yet another student can be asked to tell what one period is for the graph on the board. Then students can work on problem **21** in pairs. If time is a concern, you may want to skip this problem, or you may prefer to have students work on the Extension instead. Problem **21** may also be used as informal assessment. After students complete Section B, you may assign appropriate activities from the Try This! section, located on pages 54–57 of the *Ups and Downs* Student Book, for homework.

Comments about the Problems

21. Informal Assessment This problem assesses students' ability to identify characteristics of periodic graphs.

In order to reconstruct the shape of the track, students must reason about the relationship between the shape of the graph and the shape of the track. Students can make this problem really hard for each other. Have students decide together whether or not the reconstructed track is correctly drawn, although the shape may differ somewhat from the original drawing.

Extension Challenge students to draw another periodic graph (using the same racetrack design as in problem **21**) showing three laps in which the following events occurred:

Lap One: The race car got a flat tire halfway into the first straightaway, causing it to slow down to a crawl, make a pit stop to change the flat tire, and reenter the race at the beginning of the first curve.

Lap Two: The race car encountered no problems during this lap.

Lap Three: The race car did not slow down enough as it approached the second curve, causing it to spin out of control. The car came to a complete stop, then proceeded to continue the race.

Work Students Do

Students investigate situations that show equal increases over equal time periods in the contexts of body temperature and the thickness of a tree trunk. They make tables and graphs for these situations and see that the graphs are straight lines. Then students are introduced to the name for this type of growth: *linear growth.* They learn one way of finding the area of a circle and use the formula for the area of a circle to investigate the growth of the area of the cross section of a growing tree. In the contexts of hair growth, nail growth, and changes in rental prices, students examine how linear growth arises; how it can be represented in a table, in a graph, and in recursive and direct formulas; and how different growth rates affect the graphs and formulas.

Goals

Students will:

- use information about increase and/or decrease to create line graphs;
- identify linear patterns in tables and graphs;
- describe linear growth with recursive formulas;
- describe linear growth with direct formulas;
- make connections between situation, graph, and table;
- reason about situations of growth in terms of slope, maximum and minimum, range, decrease, and increase;
- recognize the power of graphs and/or tables for representing and solving problems.

Pacing

- approximately five 45-minute class sessions

Vocabulary

- linear growth

About the Mathematics

In this section, the concept of linear growth is formally introduced. Within the context of the given situations, the growth patterns show equal increases in equal time intervals, or a constant rate of increase. The graphs of these situations are all straight lines, which explains the name for this type of growth: linear.

As in previous sections, strong connections are made between a given situation, a table of values, a graph, and a formula. Two different types of formulas are reintroduced to describe linear relationships: recursive formulas, such as NEXT = CURRENT + 5, and direct formulas, such as $L = 2 + 5T.$ Building on students' previous experience using these formulas, students study the role of the specific numbers chosen to replace the variables and their impact on the resulting table of values and graphs. For example, within the context of comparing motorcycle rental costs for different companies, students graph and compare the formulas $P = 0.75M$ and $P = 60 + 0.20M,$ where P is the price or cost in dollars and M is the number of miles traveled, to determine which company offers the best price. Recursive and direct formulas play an important role in the unit *Growth.* Direct formulas for linear situations are further investigated in the unit *Graphing Equations.*

Materials

- Student Activity Sheets 7 and 8, pages 149 and 150 of the Teacher Guide (one of each per student)
- compasses, page 71 of the Teacher Guide (one per student)
- calculators, page 73 of the Teacher Guide (one per student)
- overhead projector, page 73 of the Teacher Guide, optional (one per class)
- graph paper, page 81 of the Teacher Guide (one sheet per student)
- colored pencils, pages 81 and 87 of the Teacher Guide (one box per student)
- paper strips, page 83 of the Teacher Guide, optional (one or two per student)
- string, page 83 of the Teacher Guide, optional (one piece per student)

Planning Instruction

You might begin this section with a brief discussion of the growth situations students encountered in Section A to ensure that they understand the way they described and identified the patterns of increase and decrease by looking at the differences in growth. Make sure that students also understand the difference between the tables of values in this section and ratio tables; not all tables are ratio tables (See the About the Mathematics section on page 13 of the Teacher Guide).

You might want students to work on problem 1 as a whole-class activity, and individually on problems 11, 14–18, 25–29, and 35–43. They may work in any setting on problems 6–10. They may work on the remaining problems in pairs or in small groups.

Problems 28, 29, and 38–41 are optional. If time is a concern, you may omit these problems or assign them as homework.

Homework

Problems 6–9 (page 70 of the Teacher Guide), 14–17 (page 74 of the Teacher Guide), and 28 and 29 (page 80 of the Teacher Guide) can be assigned as homework. After students complete Section C, you may assign appropriate activities from the Try This! Section, located on pages 54–57 of the *Ups and Downs* Student Book. The Try This! activities reinforce the key mathematical concepts introduced in this section.

Planning Assessment

- Problems 18, 42b, and 43b may be used to informally assess students' ability to describe linear growth with recursive formulas.
- Problem 25 may be used to informally assess students' ability to reason about situations of growth in terms of slope, maximum and minimum, range, decrease, and increase.
- Problem 26 may be used to informally assess students' ability to use information about increase and/or decrease to create line graphs.
- Problems 27, 42c, and 43c may be used to informally assess students' ability to describe linear growth with direct formulas.
- Problems 37 and 38 may be used to informally assess students' ability to recognize the power of graphs and/or tables for representing and solving problems.
- Problem 42a may be used to informally assess students' ability to identify linear patterns in tables and graphs.
- Problem 43a may be used to informally assess students' ability to make connections between situation, graph, and table.

C. LINEAR PATTERNS

The Marathon

In 490 B.C., there was a battle near the village of Marathon between the Greeks and the Persians. Legend has it that immediately after the Greeks won, a Greek soldier was sent from Marathon to Athens to tell the city the good news. He ran the entire 40 kilometers. When he arrived, he was barely able to stammer out the news before he died.

1. What might have caused the soldier's death?

Marathon runners need lots of energy to run long distances. Your body gets energy to run by burning food. Just like in the engine of a car, burning fuel generates heat. Your body must release some of this heat, or it will be seriously injured.

Joan Benoit won the women's marathon during the 1984 Olympics. She finished the race in 2 hours, 24 minutes, and 52 seconds. The marathon today is 42.195 kilometers long, not the original 40.

1. Answers will vary. Some students may say that the
   soldier died from heat exhaustion or a heart attack.

Overview Students read and discuss the history
of the marathon run.

Planning You might begin this section by reading
and discussing the text and completing problem **1**
as a whole-class activity. You might also ask
students how far 40 kilometers is (about 25 miles)
and how long it might take someone to run it.
(Answers will vary. If someone ran at an average
speed of 10 kilometers per hour, it would take
him or her 4 hours to run this distance.) You may
also briefly discuss how a person's body
temperature rises as a result of strenuous
exercise, such as running. This will help to
introduce the context on the next page.

Comments about the Problems

1. The purpose of this introductory question is
   to get students thinking about the fact that
   the body gets its energy from burning food
   calories and that heat is one of the by-
   products of converting food into energy.

Interdisciplinary Connection You might ask
science or health teachers in your grade level to
teach a unit on nutrition, emphasizing the impact
that food and exercise have on a person's
metabolism and energy level.

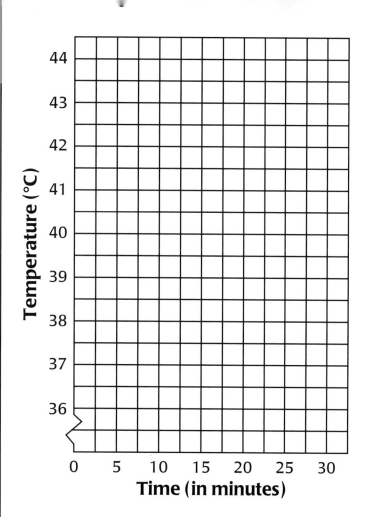

Temperature (°C)

Time (in minutes)

Normal body temperature is 37°C (98.6°F). At a temperature of 41°C (105.8°F), the body's cells stop growing. At temperatures above 42°C (107.6°F), the brain, kidneys, and other organs suffer permanent damage.

When you run a marathon, your body produces enough heat to cause an increase in body temperature of 0.17°C every minute.

2. **a.** Make a table showing how your body temperature would rise while running a marathon if you did nothing to cool off. Show temperatures every 10 minutes.

   b. Make a line graph of this data on **Student Activity Sheet 7.**

3. **a.** Why is your line graph for problem **2b** not realistic?

   b. What does your body do to compensate for the rising temperature?

When the body temperatures of marathon runners rise by about 1°C, their bodies begin to sweat to prevent the temperature from rising further. Then the body temperature neither increases nor decreases, but the body will lose about $\frac{1}{5}$ liter of water every 10 minutes.

4. Use this information to redraw the line graph from problem **2b** on **Student Activity Sheet 7.**

5. How much water do you think Joan Benoit lost during the women's marathon of the 1984 Olympics?

2. a.

Time (in minutes)	10	20	30	40
Rise in Body Temperature (°C)	1.7	3.4	5.1	6.8

b.

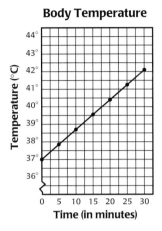

3. a. Answers will vary. Sample response:

> The line graph shows a body temperature of 42°C after 30 minutes of running, which is too high a temperature for the body to function well.

b. Your body sweats, which helps to lower the body temperature.

4.

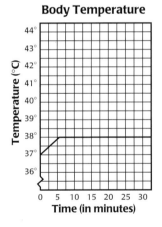

5. Answers will vary. Accept answers in the range of 2.8–3.0 liters of water. Sample strategy using a ratio table:

× 14

Time (in minutes)	10	140
Water (in liters)	$\frac{1}{5}$	$\frac{14}{5} = 2\frac{4}{5}$

Materials Student Activity Sheet 7 (one per student)

Overview Students make a table and a graph that show a marathon runner's body temperature over time if his or her temperature keeps rising by 0.17°C per minute. Then they redraw the graph taking into account one way the body cools itself.

About the Mathematics In Section A, students learned about patterns of increase and decrease by looking at the differences in the growth of one thing over equal time periods. In this section, situations that have equal increases over equal time periods are investigated. The graphs for these situations are all straight lines. The fact that there are equal increases, or a constant rate of growth, plays an important role in making formulas for these situations.

Planning Students may work on problems **2–5** in pairs or in small groups. You may wish to discuss problems **2** and **3** before students continue with problems **4** and **5.**

Comments about the Problems

2. a. It is not necessary that students make a table for the whole period of the marathon. A table that goes up to 30 minutes will suffice.

b. Make sure that every student's graph is a straight line. If students have difficulty determining where to plot the first point on the graph, you may tell them that the normal body temperature — 98.6°F — is about 37°C.

5. If students have difficulty starting this problem, you might suggest that they make a ratio table showing the amount of time in minutes and the amount of water lost in liters. Students can then use the ratio table to determine the total water Joan Benoit lost during the race.

Because the human body does not begin to sweat until the body temperature rises 1°C, which will take about 6 minutes in this situation, students need to subtract 6 minutes from the total running time to determine the amount of water lost during the race.

What's Next?

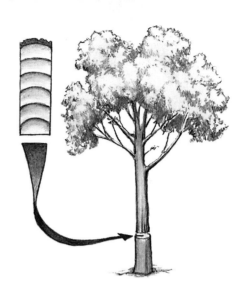

Here you see a core sample of a tree. When this sample was taken, the tree was six years old.

6. How can you tell that the tree grew steadily?

7. Use the core sample to draw a full cross section of the tree.

8. a. Make a table like the one below, showing the radius for each year.

Year	1	2	3	4	5	6
Radius (in millimeters)						

   b. Make a graph from the table.

The graph you made is a straight line. Whenever a graph is a straight line, we say that there is *linear growth* (or that the tree grew linearly).

9. a. What might have been the size of the radius in year 7? Explain how you found your answer.

   b. Suppose the tree kept growing in this way. One year the radius would be 44 millimeters. What would the radius be one year later?

If you know the radius of the tree in a certain year, you can always find the radius of the tree in the year that follows if it keeps growing linearly. In other words, if you know the radius of the CURRENT year, you can find the radius of the NEXT year.

10. If this tree continued growing linearly, how could you find the radius of the NEXT year from the radius of any CURRENT year?

6. The rings are all the same distance apart in the drawing, 4 millimeters.

7.

8. a.

Year	1	2	3	4	5	6
Radius (mm)	4	8	12	16	20	24

b.

Tree Growth

9. a. 28 millimeters. The radius increases by 4 millimeters each year. The radius of the tree was 24 millimeters during year 6. So, 24 + 4 = 28 millimeters.

b. 48 millimeters

10. Answers will vary. Sample responses:

If you add 4 to the radius of the CURRENT year, you can find the radius of NEXT year.
A formula for the radius of the tree in succeeding years is:

NEXT = CURRENT + 4, or
CURRENT + 4 = NEXT

Materials compasses (one per student)

Overview Students investigate linear growth by examining a core sample of a tree that shows rings with equal thicknesses. Students make a cross section, a table, a graph, and a formula to represent the yearly growth of the radius of the tree.

About the Mathematics On this page, students are informally introduced to recursive formulas, such as CURRENT + 4 = NEXT. The relationship between CURRENT and NEXT can be found by investigating the pattern in the rows of a table as shown below:

Years	1	2	3	4
Radius (mm)	4	8	12	16

+ 4   + 4   + 4

This relationship can also be expressed as a word formula: You can find the size of the radius for NEXT year by adding 4 millimeters to the radius for the CURRENT year. An arrow string can also be used:

$$\text{CURRENT} \xrightarrow{+ 4 \text{ mm}} \text{NEXT}$$

Planning Students may work on problems **6–10** in any setting. These problems may also be assigned as homework. Briefly discuss students' solutions and strategies, focusing on problems with which they had difficulty.

Comments about the Problems

6–9. Homework These problems may be assigned as homework.

8. b. Make sure that students connect the plotted points to make one continuous line. This is acceptable because growth is a continuous process.

9. This problem sets the stage for problem **10**. Although students can easily find the answer for part **a** by extending the given table, this strategy is rather time-consuming for part **b**. At this point, some students may see a need for a more efficient method to find a solution, such as the recursive NEXT-THEN formula, made explicit in problem **10**.

10. This is the first problem in which students are asked to express a pattern using either a word formula or an arrow string. If students have difficulty seeing the relationship here, you may need to provide additional problems.

Here you see the cross sections of two more trees. You can make graphs showing the yearly radii for these trees, too.

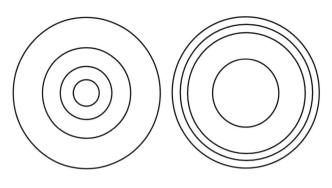

11. a. Will the graphs have straight lines or not? How can you tell without drawing the graphs?

   b. Describe the shape of the graph for each tree.

People who use wood for timber are more interested in the area of the cross section than in the size of the radius. The following series of pictures shows one way of finding the area of a circle if you know the radius.

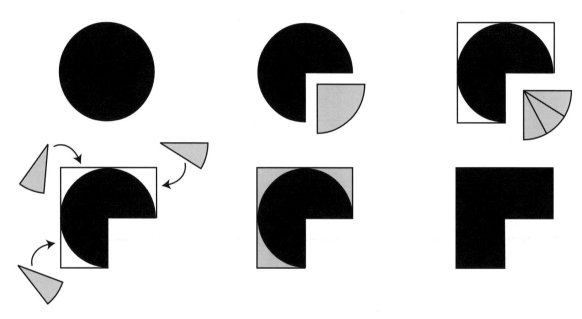

12. a. Using these pictures you can show that if a circle has a radius of five units, the area is about 75 square units. Explain how.

   b. Describe how you can find the area of a circle if you know the radius.

To find the area of a circle, you can use the general rule you found in the previous problem. A more accurate way is with the following formula:

Area of a Circle = 3.14 × *r* × *r*, where *r* is the radius of the circle

13. a. Make a table showing the area for each year of the tree discussed on the previous page.

   b. Does the area of the cross section show linear growth? Explain your answer.

11. a. No. Explanations will vary. Sample explanation:

The rings are not equally spaced, so neither radius is growing steadily (increasing at the same rate).

b. Descriptions may vary. Sample response:

First tree: the graph will become steeper and steeper, because the radius is growing faster and faster.

Second tree: the graph will become less steep and level off, because the radius grows quickly during the first two years, and then it slows down and grows at a constant rate.

12. a. The radius is the length of the side of a small square. Since each square has sides of 5 units, they each have areas of 25 square units ($5 \times 5 = 25$ square units). Since there are 3 squares (that represent the area of the one circle), the area of the circle is about 75 square units ($3 \times 25 = 75$ square units).

b. If you know the radius of the circle, you can find the area by multiplying the radius by itself and then multiplying that product by three.

13. a.

Year	1	2	3	4	5	6
Radius (mm)	4	8	12	16	20	24
Area (mm²)	50	201	452	804	1,256	1,809

b. No. Explanations will vary. Sample explanations:

• The area doesn't increase by the same amount every year.

• The growth of the area is not the same every year.

• The differences in the table are not equal.

Materials calculators (one per student); overhead projector, optional (one per class)

Overview Students examine cross sections of two different trees to determine whether or not the graphs that show their annual growth will be straight lines. Then they investigate whether or not the area of a given cross section grows linearly.

About the Mathematics The conventional formula for the area of a circle is often better understood when the circle's area is visually realloted to make three squares, each with sides equal to the circle's radius, as shown on Student Book page 32. Students may recall this strategy from the unit *Reallotment*. An alternative method for finding the formula for the area of a circle is presented in the unit *Going the Distance*. Students are also formally introduced to π (pi) in that unit.

Planning Students may work individually on problem **11** and in pairs or in small groups on the remaining problems. After students finish problem **11,** you might want to read and discuss the text and diagrams about how to reallot the area of a circle together with students. Briefly review students' solutions and strategies for these problems, focusing especially on problems **12b** and **13.**

Comments about the Problems

11. Writing Opportunity You might have students write their answers to this problem in their journals.

12. b. Make a drawing of this circle on the board or overhead projector and demonstrate how the circle's area can be realloted to make three squares. You may need to explain how the picture of the three squares can be represented using the formula $A = 3 \times r \times r$. The 3 refers to the 3 squares, and $r \times r$ refers to the area of each square. Since the circle's wedges do not make an exact fit in the spaces, the formula $A = 3 \times r \times r$ is an approximation of the conventional formula, $A = \pi r \times r$.

13. b. If students are having difficulty, you may suggest that they make a graph.

Hair and Nails

Paul went to get a haircut. When he got home, he looked in the mirror and screamed, "It's too short!"

He decided not to get his hair cut again for a long time. In the meantime, he decided to measure how fast his hair grew. Below is a table that shows the length of Paul's hair (in centimeters) as he measured it each month.

Time (in months)	0	1	2	3	4	5	6
Length (in centimeters)	2	3.5	5	6.5			

14. How long was Paul's hair after the haircut?

15. a. How long will his hair be in five months?

    b. Why is it easy to calculate this length?

16. How long will Paul's hair be after a year if it keeps growing at the same rate and he does not get a haircut?

17. If Paul's hair is 10 centimeters long at some point, how long will it be one month later?

14. 2 centimeters

15. a. 9.5 centimeters

b. Explanations will vary. Sample explanation:

It is easy to calculate the new hair length because his hair grows the same amount (1.5 cm) every month.

16. 20 centimeters. Strategies will vary. Sample strategies:

• Some students might extend the table and fill in lengths for the remaining six months.

Time (in months)	0	6	7	8	9	10	11	12
Length (in centimeters)	2	11	12.5	14	15.5	17	18.5	20

+ 3   + 3

• In 12 months, Paul's hair grew 18 centimeters (12 × 1.5 = 18 cm). His hair was originally 2 centimeters long, so after one year, his hair length will be 2 + 18 = 20 centimeters long.

• Paul's hair grew 9 centimeters in 6 months. So, during the next 6 months, it will grow another 9 centimeters. Altogether, that will make his total hair length 20 centimeters (11 + 9 = 20 cm).

17. 11.5 centimeters

Overview Students investigate a person's hair growth using the data in a table. They make predictions about what his hair length would be in five months.

About the Mathematics A table is a strong tool to investigate growth patterns. Students may now feel comfortable with the concept of growth rate. In this table, the growth rate is 1.5 centimeters per month. Because the differences in length for every two consecutive months are equal, this is a linear growth pattern.

Planning Students may work individually on problems **14–17.** You may assign problems **14–17** as homework.

Comments about the Problems

14–17. Homework These problems may be assigned as homework. The purpose of these problems is to help students to begin thinking about how to write a formula to represent this hair-growth situation. They write the formula in problem **18** on the next Student Book page.

16. Encourage students to show or explain how they got their answers. Students who have answers of 22 or 26 centimeters may have mistakenly used the given table as a ratio table. Distinguish the table of values here from a ratio table with these students. (See the About the Mathematics section on page 31 of this Teacher Guide.).

17. By this point in the unit, students should be able to understand the relationship between the time and hair-length values in the table and use this relationship to predict future growth. After students determine the given relationship here, you may expect that they will have little trouble in expressing this growth pattern using a word formula or an arrow string. Again, be sure that students show their work or explain their reasoning.

18. If you know the length of Paul's hair in the CURRENT month, you can use it to find his hair length for the NEXT month. Write a formula using NEXT and CURRENT.

19. Draw a graph showing how Paul's hair grows over a year if he does not get a haircut. Describe the shape of this graph.

Sonya's hair grew about 14.4 centimeters in one year. It is possible to write the following formulas:

$$NEXT = CURRENT + 14.4$$

$$NEXT = CURRENT + 1.2$$

20. Explain what each formula represents.

18. NEXT = CURRENT + 1.5, or
CURRENT + 1.5 = NEXT

19.

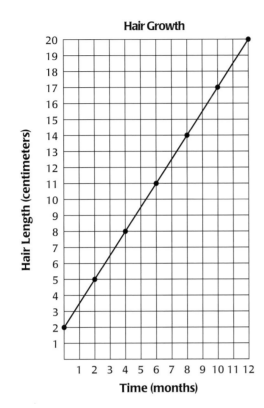

Hair Growth

Paul's hair is growing by equal amounts each month, so the graph will be a straight line.

20. The first formula gives Sonya's hair growth each year, so NEXT stands for next year, CURRENT for the current year, and 14.4 stands for the number of centimeters her hair grows yearly.

The second formula gives her hair growth each month, so NEXT stands for next month, CURRENT stands for the current month, and 1.2 stands for the number of centimeters her hair grows monthly.

Overview Students write a formula to represent the hair growth using NEXT and CURRENT. Then they explain two other NEXT/CURRENT formulas.

Planning You may want to have students work individually on problem **18,** which can be used as an informal assessment. Discuss the notation used in the formula after students complete this problem. Then students may continue working in small groups on problems **19** and **20.**

Comments about the Problems

18. Informal Assessment This problem assesses students' ability to describe linear growth with recursive formulas.

If students have difficulty writing the NEXT/CURRENT formula, you may refer them to problem **17.** You might also help students to write a word formula first, by asking: *How can you find Paul's hair length for the next month?* [You can find the length of his hair the next month by adding 1.5 centimeters to the current length.] Most students will be able to then translate this word formula into the NEXT-THEN formula.

Lastly, you could also suggest that students express the growth pattern using an arrow string. Give full credit if students show an understanding of the growth pattern using either a word formula or an arrow string.

20. The purpose of this problem is to help students to realize the importance of knowing what the words and/or symbols used in a formula represent.

21. Make a graph that might show the length of Sonya's hair over one year's time if she goes to the hairdresser every two months. Be prepared to explain your graph.

The formula for Sonya's hair growth does not include information about how long her hair was at the beginning of the year. You knew that the beginning length of Paul's hair was 2 centimeters. That's why it is possible to make another formula for Paul's hair growth:

$$L = 2 + 1.5T$$

22. What do you think the letter L stands for? the letter T? Explain the numbers in the formula.

23. Sacha's hair is 20 centimeters long and grows at a constant rate of 1.4 centimeters a month. Write a formula with L and T to describe the growth of Sacha's hair.

24. Look at the graph you made for Sonya's hair growth, but look only at the part that shows the growth in the first two months. Use this information to write a formula using L and T for Sonya's hair growth.

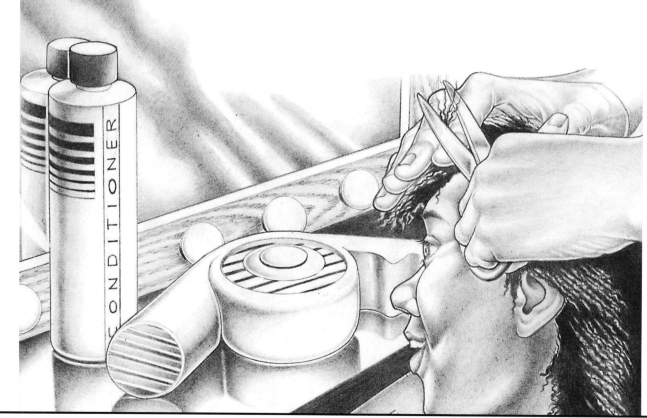

21. Answers will vary. Sample response:

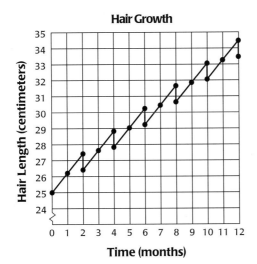

Hair Growth

Sonya's hair is initially 25 centimeters long. Sonya likes to wear her hair long and has only 1 centimeter trimmed off each time.

22. *L* stands for length, in centimeters.
T stands for time, in months.
The number 2 is the beginning hair length, 2 centimeters.
The number 1.5 stands for the amount it grows each month, 1.5 centimeters.

23. Formulas will vary. Sample formulas:

$L = 20 + 1.4T$, or $L = 20 + 16.8T$, or
$20 + 1.4 \times T = L$ $20 + 16.8T = L$

24. Answers will vary, depending on individual student graphs made from problem **21.** If students chose 25 centimeters as the starting length for Sonya's hair, the formula would be $L = 25 + 1.2T$.

Overview Students make a graph to show hair length over a year's time. Then they explain a direct formula that describes hair growth. They write this type of formula to describe the hair growth of another person.

About the Mathematics Formulas like $L = 2 + 1.5T$ are called *direct* formulas. All linear growth situations can be described with a direct formula similar to this one. Formulas and linear equations are studied more extensively in the unit *Graphing Equations*.

Planning Students may continue working in small groups on problems **21–24.** If you notice students having difficulties with problem **22,** you may want to stop here and have a class discussion about the meanings of the letters and numbers in the formula.

Comments about the Problems

21. Students should realize they not only must choose a beginning hair length themselves, but also determine how much is cut off each time. (See the Hints and Comments column for problem **26** on Teacher Guide page 81.)

22. By this point in the unit, most students should understand what each letter in the formula represents and how the formula originated. If students are having difficulty, you might again suggest that they write a word formula to represent this pattern:

You can find the hair length for a given month by adding 1.5 centimeters to the original hair length of 2 centimeters.

Students may write the direct formula in different orders: $2 + 1.5 \times T = L$ or $2 + T \times 1.5 = L$.

23. Students may use two different growth rates in their formulas: 1.4 centimeters (the monthly rate) or 16.8 centimeters (the yearly rate). In the later case, the *T* must represent the number of years rather than the number of months.

Suppose you decided not to cut one fingernail for several months and the nail grew at a constant rate. Below you see the lengths of a nail in millimeters at the beginning and after four months.

Time (in months)	0	1	2	3	4
Fingernail Length (in millimeters)	15				23

25. How much did this nail grow every month?

26. Make a graph that shows how the length of the fingernail is changing.

27. Write a formula for fingernail growth using *L* for length and *T* for time.

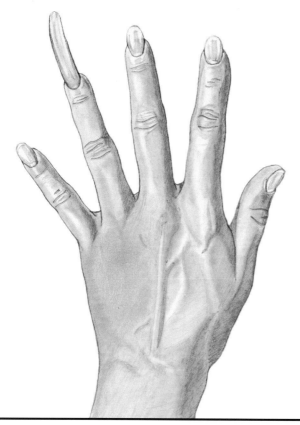

Fingernails grow about four times as fast as toenails.

28. Make a table to show how a toenail would grow if it were never cut or broken. Choose a reasonable starting length.

29. a. Make a graph that represents toenail growth on the same diagram you used for the graph of fingernail growth in problem **26.** Use a different-colored pencil so that you can tell the lines apart.

b. Compare the two linear graphs. How do they show that fingernails grow about four times as fast as toenails?

c. Write a formula for toenail growth using length (*L*) and time (*T*).

25. 2 millimeters. Strategies will vary. Some students might reason that a growth pattern of 8 millimeters over a four-month period is the same as 2 millimeters per month.

26.

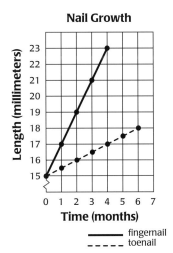

Nail Growth

--- fingernail
- - - - toenail

27. $L = 15 + 2T$

28. Answers will vary, depending on the starting length chosen, but the table values should increase by 0.5 mm each month (which is $\frac{1}{4}$ the rate of fingernail growth).

Sample table:

Time (months)	0	1	2	3	4	5	6
Toenail Length (mm)	15	15.5	16	16.5	17	17.5	18

29. a. See the toenail graph in the answer to problem **26** above.

b. The fingernail graph is steeper. If you look at the graph between 0 and 2 months, for example, the toenail has grown 1 millimeter and the fingernail has grown 4 millimeters.

c. L = (beginning toenail length) + $0.5T$

Materials graph paper (one sheet per student); pencils in two colors (one of each color per student)

Overview Students investigate the linear growth patterns of fingernails and toenails. They make tables, construct graphs, and write formulas to describe the growth pattern in each situation.

Planning Students may work on problems **25–29** individually. Problems **25–27** may be used as informal assessment. Problems **28** and **29** are optional.

Comments about the Problems

25. Informal Assessment This problem assesses students' ability to reason about growth situations in terms of slope, maximum and minimum, range, decrease, and increase.

In this problem, students must reason about the monthly growth rate, given that this situation represents a linear growth pattern. If students are having difficulty, you might suggest that they determine how many millimeters the fingernail grew between the first month and fourth month (8 mm) and remind them that the fingernail is growing at a constant rate.

26. Informal Assessment This problem assesses students' ability to use information about increase and/or decrease to create line graphs.

Students must provide the scale for the vertical axis of the graph themselves. Many students may choose to label the vertical axis counting by twos. You might suggest that they start with the number 15 to avoid having to extend the vertical axis from 1 to 23 (see the sample graph in the solutions column for this problem).

27. Informal Assessment This problem assesses students' ability to describe linear growth with direct formulas.

28–29. Homework These problems may be assigned as homework.

28. Students must first determine the growth rate of the toenails before they can make a table. Some students may use the following reasoning: if fingernails grow about four times as fast as toenails, then toenails grow at a rate that is one-quarter of the growth rate of fingernails. So, $\frac{1}{4}$ of 2 millimeters per month is $\frac{1}{2}$ millimeter per month.

Renting a Motorcycle

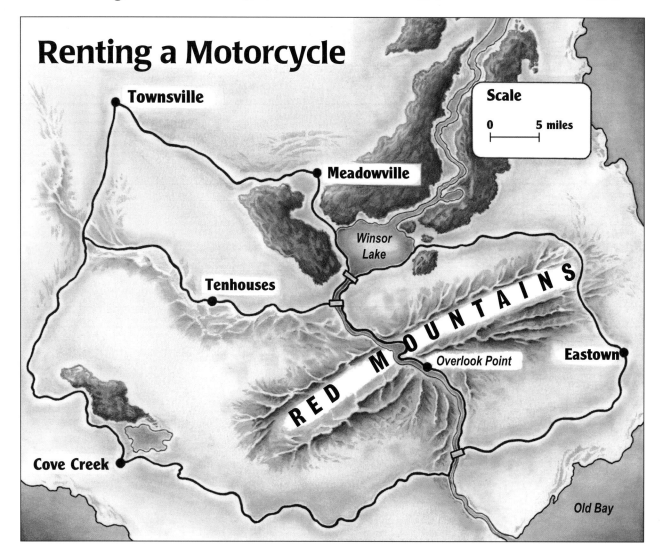

Scale

0   5 miles

Townsville

Meadowville

Winsor Lake

Tenhouses

RED MOUNTAINS

Overlook Point

Eastown

Cove Creek

Old Bay

During the summer months, many people visit Townsville. A popular tourist activity there is to rent a motorcycle and take a one-day tour through the mountains.

You can rent motorcycles at E. C. Rider Motorcycle Rental and at Budget Cycle Rental. The two companies calculate their rental prices in different ways.

The most popular trip this season goes from Townsville, through Cove Creek, to Overlook Point, and back through Meadowville.

30. Estimate the number of miles in this trip.

MOTORCYCLE RENTAL

One Day: $60
Plus $0.20 per Mile

CYCLE RENTAL

One Day:
Just $0.75 per Mile

30. Estimates will vary, but the trip is about 170 miles. Accept answers in the range of 150–180 miles.

Materials paper strips, optional (one or two per student); string, optional (one piece per student)

Overview Students estimate the distance of a motorcycle trip on a map. This distance is important for the problems on Student Book pages 38 and 39.

About the Mathematics The remaining problems in this section will reinforce students' understanding of linear growth patterns and formulas. These problems also will provide students with enough experiences to begin to understand the concepts of intercept and slope, which are made explicit in the unit *Graphing Equations*.

Planning Students may work on problem **30** in small groups.

Comments about the Problems

30. Students might use string or a paper strip to estimate the mileage for this trip.

Even though more and more people are making this trip, the owner of Budget Cycle Rental notices that her business is getting worse and worse. This is very surprising to her, because her motorcycles are of very good quality.

31. Explain the decrease in Budget's business.

The rental price you pay depends on the number of miles you ride. With Budget Cycle Rental, the price goes up $0.75 for every mile you ride.

32. a. How much does it go up per mile with a rental from E. C. Rider?

    b. Does that mean that it is less expensive to rent from E. C. Rider? Explain your answer.

Budget Cycle Rental uses the rental formula: $P = 0.75M$.

33. a. Explain each part of this formula.

    b. What formula can E. C. Rider use?

34. Graph both formulas on **Student Activity Sheet 8.**

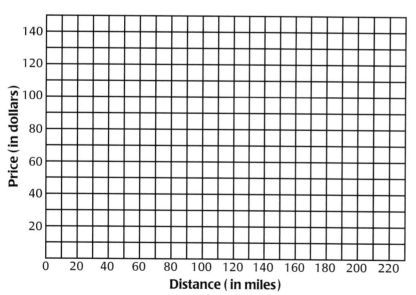

31. Most people rent where it is the least expensive.

E. C. Rider's charge for the trip would be
$60.00 + 170 × $0.20 = $94.00.

Budget's charge for the trip would be
170 × $0.75 = $127.50.

It is less expensive to rent from E. C. Rider for this trip.

32. a. 20 cents per mile

b. No. Explanations will vary. Sample explanation:

The better choice depends on the length of the trip. To rent a motorcycle from E. C. Rider for a 10-mile trip would cost $60 + 10 × $0.20 = $62.00. To rent from Budget for the same distance would cost 10 × $0.75 = $7.50.

33. a. *P* represents the price or cost in dollars. *M* represents the number of miles of the trip. The price you pay per mile is $0.75.

b. $P = 60 + 0.20M$

34.

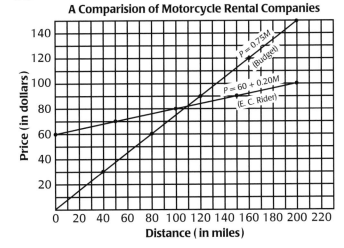

A Comparision of Motorcycle Rental Companies

Materials Student Activity Sheet 8 (one per student)

Overview Students compare motorcycle rental prices of two companies. They write formulas to represent the rental pricing structure of both companies. Then they draw a graph to compare the two companies' rental prices.

About the Mathematics These problems involve linear growth patterns. Students should see the similarities between these situations and the situations of hair and nail growth: they all deal with equal increases over equal time periods. The hair growth situation may be used as a model to show students an example of linear increase.

E. C. Rider charges $0.20 per mile. This is the amount of growth per month. This number determines the steepness of the graph. E. C. Rider's charges start with an amount of $60. This is analogous to the beginning length in the hair-growth context. This number can be found where the graph starts on the vertical axis.

Planning Students may continue to work on problems **31–34** in small groups. Encourage students to discuss their solutions together before they begin each new problem. If possible, wait until students have finished problem **36** on Student Book page 39 to have a class discussion.

Comments about the Problems

31. Students must realize that they can use the distance from problem **30** to calculate the price that each company charges. Although students' answers to problem **30** may differ, they will reach the same conclusion here.

32. b. Students may have already noticed that a short trip would be more expensive at E. C. Rider because of that company's fixed charge of $60.

33. Students should be able to write the 0.20*M* part of the formula.

34. Most students will need to make tables before they will be able to draw a graph of both formulas. Let students struggle for a short time before you suggest making a table. Hopefully, they will come up with the idea themselves.

35. If someone asked you where to rent a motorcycle in Townsville, what would you advise and why?

36. The trip most people make is about 170 miles. Look at the two graphs from problem **34** and explain why most people rent from E. C. Rider.

Ms. Rider is thinking about changing her formula. She thinks about raising the starting amount from $60 to $70.

37. a. What would the new formula be?

    b. Make a graph of this new formula in the diagram you already have.

    c. Do you think Ms. Rider's idea is a good one? Why or why not?

Ms. Rider has another idea. She proposes raising the price 0.10¢ per mile and keeping the starting amount at $60.

38. a. What formula can you write for this proposal?

    b. Graph the formula on **Student Activity Sheet 8** and use the graph to decide if this idea is a good one.

Now a real competition between the two motorcycle companies has begun. Budget Cycle Rental is going to lower prices. See the new sign to the left.

39. Write the new formula for Budget Cycle Rental.

40. Make a graph of this new formula on **Student Activity Sheet 8.**

41. Take another look at the 170-mile trip from Townsville. Who would you rent your motorcycle from now, given the new information from problems **38** and **39?**

BUDGET CYCLE RENTAL

One Day:
Just $0.75 per Mile

THE FIRST 20 MILES

FREE!

35. Answers will vary. Sample response:

I would tell them that it depends on the total miles of their trip. If their trip was more than about 110 miles, then E. C. Rider would be cheaper.

36. Explanations will vary. Sample explanation:

You can see from the graph that at 170 miles, the E. C. Rider graph is below the Budget graph, so E. C. Rider's price is less expensive.

37. a. $P = 70 + 0.20M$

b.

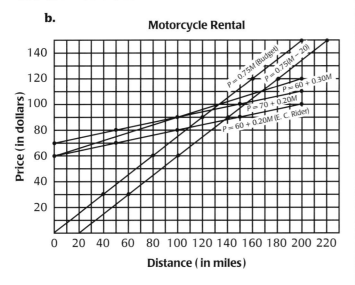

Motorcycle Rental

Price (in dollars) vs *Distance (in miles)*

Graph lines labeled:
$P = 0.75M$ (Budget)
$P = 0.75(M - 20)$
$P = 60 + 0.30M$
$P = 70 + 0.20M$
$P = 60 + 0.20M$ (E. C. Rider)

c. Answers may vary. Sample response:

Ms. Rider's price will still be cheaper than Budget's price for the most popular trip, so I think it is a good idea.

38. a. $P = 60 + 0.30M$

b. See the graph in the solution to problem **37b.**

Ms. Rider's price for the most popular trip is still cheaper than Budget's price, and on longer trips, she will make more money. So, I think this is a good idea.

39. $P = 0.75 (M - 20)$, or $(M - 20) \times 0.75 = P$

40. See the graph in the solution to problem **37b.**

41. E. C. Rider's price ($111.00) is slightly cheaper than Budget's price ($112.50).

Materials colored pencils (one box per student)

Overview Students analyze the graphs of the two companies' rental formulas to compare their costs for a 170-mile trip. They then rewrite the formulas to reflect a new price structure and draw graphs to analyze the effects of a price war between the two companies.

About the Mathematics As students draw their graphs in problems **37, 38,** and **40,** they might notice that the starting point for the new graph line for each company's rental formula is translated. Also, the points of intersection (that indicate trips for which the price is the same for either company) are different. The new graph lines still have the same steepness (slope).

Planning After students complete problem **36,** discuss problems **31–36** with the whole class. Problems **37** and **38** may be used as informal assessment. Problems **38–41** are optional. If time is a concern, you may omit these problems or assign them as homework. You might want students to work individually on problems **35–41.** Be sure that students use different colored pencils to indicate different graphs and write the formula's name on each graph in problems **37b, 38b,** and **40.**

Comments about the Problems

35. Ask students how they might use their graph from problem **34** to answer this problem. The graph shows that the costs of both companies are the same for a trip of about 110 miles. For trips less than 110 miles, Budget is less expensive, and for trips more than 110 miles, E. C. Rider is less expensive.

37–38. Informal Assessment These problems assess students' ability to recognize the power of graphs and/or tables for representing and solving problems.

38. Be sure that students understand that the price per mile has been raised by 10 cents, so that the $0.20M$ changes into $0.30M$ in the formula.

39. This may be a difficult problem for many students, since the graph for Budget starts at 20 on the horizontal axis.

Summary

The situations in this section were all examples of graphs with straight lines. A graph with a straight line shows linear change.

You can recognize linear change by looking at the numbers in a table or by considering a graph. If something is changing linearly, the differences over equal time periods will always be the same.

Summary Questions

Time (in months)	0	1	2	3
Length (in centimeters)	11	12.4	13.8	15.2

42. a. Using the table above, show that the growth occurring is linear.

b. Write a formula using NEXT and CURRENT for the above example.

c. Write a formula using L (length) and T (time) for the above example.

43. Samantha earns $12 per week by baby-sitting.

a. Make a table to show how much money Samantha earns over six weeks.

b. Write a formula using NEXT and CURRENT to describe Samantha's earnings.

c. Write a formula using W (week) and E (earnings) to describe Samantha's earnings.

42. a. The table shows that the length grows each month by the same amount; the differences are all equal to 1.4 centimeters.

b. NEXT = CURRENT + 1.4

c. $L = 11 + 1.4T$

43. a.

Time (weeks)	0	1	2	3	4	5	6
Earnings (dollars)	0	12	24	36	48	60	72

b. NEXT = CURRENT + 12

c. $E = 12W$

Overview Students summarize what they have learned about linear growth by making tables, formulas, and graphs of linear situations in the Summary problems.

Planning Have students first read and discuss the Summary. Students may work individually on problems **42** and **43** to see how well each student understands linear situations. Problems **42** and **43** can be used as informal assessment. After students complete Section C, you may assign appropriate activities in the Try This! section, located on pages 54–57 of the Student Book, for homework.

Comments about the Problems

42. Informal Assessment Problem **42a** assesses students' ability to identify linear patterns in tables and graphs.

Problem **42b** assesses students' ability to describe linear growth with recursive formulas.

Problem **42c** assesses students' ability to describe linear growth with direct formulas.

43. Informal Assessment Problem **43a** assesses students' ability to make connections between situation, graph, and table.

Problem **43b** assesses students' ability to describe linear growth with recursive formulas.

Problem **43c** informally assesses students' ability to describe linear growth with direct formulas.

Note that this problem, like problem **42,** does not describe a continuous situation; therefore the dots in the graph should not be connected. Even though students are not asked to draw the graph, you might want to discuss this point.

Work Students Do

Students continue to explore growth factors in this section. They discover that not all growth is linear through their examination of how aquatic weeds in Africa, water lilies in South America, and bacteria grow over time. Students use tables, graphs, and recursive formulas to recognize when growth increases faster and faster, or exponentially, over time.

Goals

Students will:

- identify and describe patterns of increase and/or decrease from a table or graph;
- understand and use the concept of growth factor;
- identify a growth factor;
- recognize the power of graphs and/or tables for representing and solving problems;
- use algebraic models to represent realistic situations;
- use information about increase and/or decrease to create line graphs.*

  * *This goal is addressed in this section and assessed in other sections in the unit.*

Pacing

- approximately two 45-minute class sessions

Vocabulary

- growth factor

About the Mathematics

In this section, a pattern of increase known as *exponential growth* is informally introduced to students. Exponential growth can be modeled with repeated multiplication. This contrasts with linear growth, which can be modeled with repeated addition. Exponential growth is a function in which the growth factor is a number greater than one. When the growth factor is a number between zero and one, the function is known as *exponential decay.* This concept is examined in Section E.

Students are introduced to the concept of doubling, a special type of exponential growth, in *Patterns and Symbols* when they investigated super-even numbers (powers of 2). The concept of doubling is revisited in several other units. The conventional exponential notation is made explicit in the unit *Powers of Ten.*

Recursive formulas (such as NEXT = 2 × CURRENT) are introduced to describe exponential growth situations. Direct formulas for exponential growth (such as $A = 2^n$) are formally studied in the grade 8/9 unit *Growth.*

Materials

- Student Activity Sheets 9 and 10, pages 151 and 152 of the Teacher Guide (one of each per student)
- colored pencils, page 95 of the Teacher Guide (one box per group of students)
- graph paper, page 97 of the Teacher Guide (one sheet per student)
- centimeter rulers, page 101 of the Teacher Guide (one per student)
- calculators, page 101 of the Teacher Guide (one per student)

Planning Instruction

You might introduce the context at the beginning of this section with a short discussion about how bacteria grows. Every time a bacteria cell splits into two cells, the number of bacteria doubles. You might challenge students to answer the following related brain teaser now, or use it as an Extension activity later in this section:

There is one bacteria in a test tube. Every minute, the bacteria reproduces, doubling the number of bacteria. In other words, after one minute, there are two bacteria, after two minutes, there are four bacteria, and so on. The bacteria will totally fill the test tube after one hour. After how many minutes will the test tube be exactly half filled? [After 59 minutes, the test tube will be half filled.]

Students may work on problems 8 and 9 and 17–19 individually. Students may work on problems 6 and 7 individually or in small groups. They may work on the remaining problems in pairs or in small groups.

There are no optional problems in this section.

Homework

Problems 8 and 9 (page 96 of the Teacher Guide) may be assigned as homework. Also, the Writing Opportunity (page 103 of the Teacher Guide) may be assigned as homework. After students complete Section D, you may assign appropriate activities from the Try This! section, located on pages 54–57 of the *Ups and Downs* Student Book. The Try This! activities reinforce the key mathematical concepts introduced in this section.

Planning Assessment

- Problems 6, 17, and 18 may be used to informally assess students' ability to understand and use the concept of growth factor.
- Problem 7 may be used to informally assess students' ability to identify and describe patterns of increase and/or decrease from a table or graph and to recognize the power of graphs and/or tables for representing and solving problems.
- Problems 17 and 18 may be used to informally asses students' ability to use algebraic models to represent realistic situations.
- Problem 18 may be used to informally assess students' ability to identify a growth factor.

Aquatic Weeds

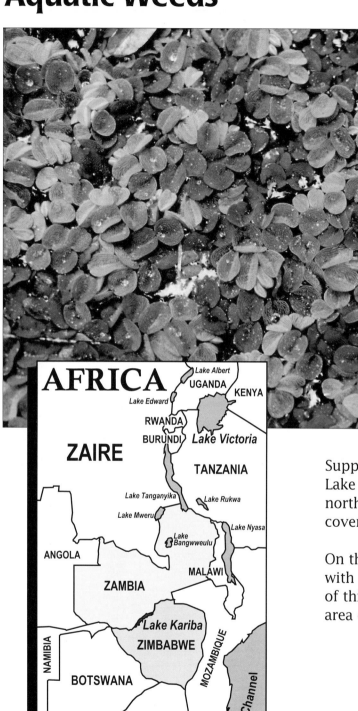

AFRICA

The waterweed *Salvinia auriculata*, found in Africa, is a fast-growing weed. In 1959, a patch of *Salvinia auriculata* was discovered in Lake Kariba on the border of what are now Zimbabwe and Zambia. People noticed it was growing very rapidly.

The first time it was measured, in 1959, it had grown to cover 199 square kilometers. A year later, it covered about 300 square kilometers. In 1963, the weed covered 1,002 square kilometers of the lake.

1. Did the area covered by the weed grow linearly? Explain.

Suppose a different weed was found in Lake Victoria, which is about 1,000 miles north of Lake Kariba, and that the area it covered doubled every year.

On the next page is a map of Lake Victoria with a grid pattern drawn on it. One square of this grid is colored in to represent the area covered by the weed in one year.

1. No. Explanations will vary. Sample responses:

- The first year the area increased by about 100 square km. If it grew linearly, it would have grown this same amount every year. But we are told that in 1963 the weed covered 1,002 km^2, so the weed did not grow linearly.

- From 1959 to 1963 (in four years), the area increased by about $1,000 - 200 = 800$ km^2. So if the area increased linearly, it would have increased by $800 \div 4 = 200$ km^2 per year. But from 1959 to 1960, the area increased from 199 to 300 km^2, and that difference is not 200 km^2.

Overview Students investigate the growth of the area in a lake that is covered by a fast-growing weed.

About the Mathematics The context of a growing weed informally introduces students to the concept of exponential growth, the topic of this section. While linear growth can be identified by looking for equal increases over equal time periods, exponential growth is identified differently. (See the About the Mathematics section on Teacher Guide page 97.) In both growth-pattern situations, equal time periods must be considered.

Planning You might want to read and discuss the text with students. They may solve problem **1** in small groups. Be sure to discuss students' solutions and strategies for this problem and read the text below problem **1.** Make sure students understand that the area covered by the weed doubles every year.

Comments about the Problems

1. Students should realize that the growth rate between the given years is not equal. You may suggest that they use a table, as shown below, to see whether or not the growth amounts can be derived from the growth rate of the area between any two given years.

Year	1959	1960	1963
Area (in km²)	199	300	1,002

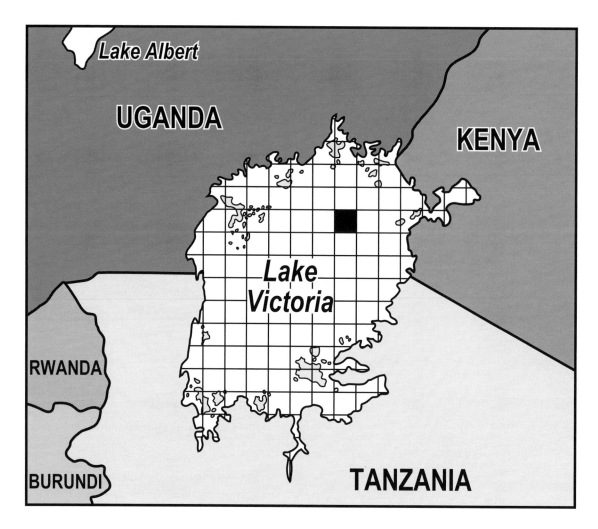

Use **Student Activity Sheet 9** to answer problems **2–5.**

2. If the shaded square represents the area currently covered by the weed, how many squares would represent the area covered next year? a year later? a year after that?

3. Angela is showing the growth of the area covered by the weed by coloring squares on the map. She uses a different color for each year. She remarks: "The number of squares I color for a certain year is exactly the same as the number already colored for all of the years before." Use **Student Activity Sheet 9** to show why Angela is or is not correct.

4. How many years would it take for the lake to be about half covered?

5. How many years would it take for the lake to be totally covered?

2. Next year: two squares
A year later: four squares
A year after that: eight squares

3. Angela is correct. Explanations will vary. Sample explanations:

In one year, there are 8 squares covered. The next year, that number will double to make 16 squares covered. So, you need to color 8 additional squares.

To double something, you add that amount to itself (2 × area = area + area).

Using the map:

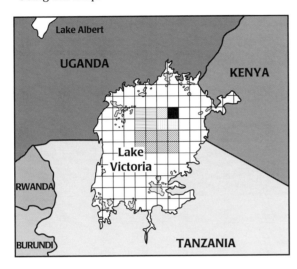

4. It would take between five and six years for the lake to be half covered.

Explanations will vary. Sample explanation:

The lake covers an area of 100 square units. After five years, 32 square units would be covered, and after six years, 64 square units would be covered. So, it would take somewhere between five and six years for the lake to be half covered.

5. It would take between six and seven years for the lake to be fully covered.

Explanations will vary. Sample explanation:

The lake covers an area of 100 square units. After six years, 64 square units would be covered, and after seven years, more than 100 square units would be covered. So, it would take somewhere between six and seven years for the lake to be fully covered.

Materials Student Activity Sheet 9 (one per student); colored pencils (one box per group)

Overview Students use a map to investigate the area of weed growth on a lake over a time span.

About the Mathematics By coloring the square units in the area, students will realize that the total number of squares they have colored at the end of the current year is equal to the number they will color next year. This may prepare students to later write a recursive formula to describe this exponential growth situation.

By the end of this section, students should understand the basis for exponential growth used in this section: doubling. They should be able to look at doubling patterns in two different ways, as shown below.

To find the area for the fifth year, take the area of the current year (8) and add the same amount: 8 + 8 = 16 square units.

Year	0	1	2	3	4
Area (squares)	1	2	4	8	16

+ 1   + 2   + 4   + 8

This leads to the formula:

NEXT = CURRENT + CURRENT

Another way to find the area for the fifth year is to take the previous year's area and multiply it by two: 4 × 2 = 16 square units.

Year	0	1	2	3	4
Area (squares)	1	2	4	8	16

× 2   × 2   × 2   × 2

This leads to the formula:

NEXT = 2 × CURRENT

A formal discussion about these recursive formulas is not necessary at this point.

Planning Students may work on problems **2–5** in pairs or in small groups. Be sure to discuss students' solutions and strategies for these problems.

Comments about the Problems

3. Some students may have difficulty with the concept of doubling. They may use the following incorrect reasoning: There are 8 squares colored this year. Double that and you get 16, so I need to color 16 new squares.

Double Trouble

Rajpreet is studying a type of bacteria at school. Bacteria usually reproduce by cell division. During cell division, a bacterium splits in half and forms two new cells. Each bacterium then splits again, and so on. These bacteria are said to have a *growth factor* of two because their amount doubles after each time period.

6. Suppose the number of cells for these bacteria doubles every 20 minutes. Starting with a single cell, calculate the number of existing bacteria after 2 hours and 40 minutes. Make a table like the one below to find your answer.

Time (in minutes)	0	20	40
Number of Bacteria			

7. a. Are the bacteria growing linearly? Explain.

   b. Graph the changes in the number of bacteria over the course of 2 hours and 40 minutes. Describe the graph.

Two thousand bacteria are growing in the corner of the kitchen sink. You decide it is time to clean house. You use a cleanser on the sink which is 99% effective in killing bacteria.

8. How many bacteria survive your cleaning?

9. If the number of bacteria doubles every 20 minutes, how long will it take before there are as many bacteria as before?

6. After 2 hours and 40 minutes, or 160 minutes, there are 256 bacteria.

Time (in minutes)	0	20	40	60	80	100	120	140	160
Number of Bacteria	1	2	4	8	16	32	64	128	256

7. a. No. The bacteria are not growing linearly. Explanations will vary. Sample explanation:

If you calculate the differences, you will see that the number of bacteria is not increasing by equal amounts:

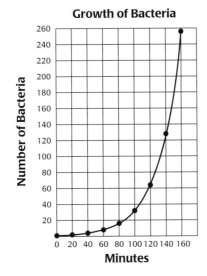

Time (in minutes)	0	20	40	60
Number of Bacteria	1	2	4	8

+ 1   + 2   + 4

b.

Growth of Bacteria

Number of Bacteria vs *Minutes*

Descriptions will vary. Sample description:

The graph is not a straight line. It becomes steeper and steeper, showing that the number is increasing faster and faster.

8. There will be 20 bacteria left. Some students may reason that 99% are killed, so 1% survive, and 1% of 2,000 is 20.

9. Between 120 and 140 minutes. Some students may use a table to find the answer. Sample table:

Time (in minutes)	0	20	40	60	80	100	120	140
Number of Bacteria	20	40	80	160	320	640	1,280	2,560

Materials graph paper (one sheet per student)

Overview Students are introduced to the concept of a growth factor. They use the given growth factor to solve problems about the growth of bacteria.

About the Mathematics Exponential growth can be modeled with repeated multiplication. In the table below, the multiplier (× 2) is the ratio between the two numbers for adjoining time periods. In exponential patterns, this multiplier is known as the *growth factor*. Some students may remember the term *factor of enlargement* from the unit *Ratios and Rates*. Use this term to help reinforce the idea that a factor deals with the operation of multiplication.

Planning Read and discuss the text on Student Book page 43 together. Make sure students understand the term *growth factor*. Students may work on problems **6** and **7** individually or in small groups. These problems may also be used as informal assessment. Problems **8** and **9** may be assigned as homework.

Comments about the Problems

6. Informal Assessment This problem assesses students' ability to understand and use the concept of growth factor. Some students may arrive at the incorrect answer of 16 by using the following table.

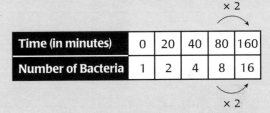

Time (in minutes)	0	20	40	80	160
Number of Bacteria	1	2	4	8	16

× 2

7. Informal Assessment This problem assesses students' ability to identify and describe patterns of increase and/or decrease from a table or graph, and their ability to recognize the power of graphs and/or tables for representing and solving problems.

Students might have difficulty in using an appropriate scale for the axes. Be sure to discuss the idea that the graph shows that doubling results in an increasing growth rate.

8–9. Homework These problems may be assigned as homework.

9. You may ask students how they might find a more precise answer to this problem. Some students might suggest using a graph to obtain a more precise answer.

Water Lily

The *Victoria regina*, named after Queen Victoria of England, is a very large water lily that grows in South America. The name was later changed to *Victoria amazonica*. The leaf of this plant can grow to nearly 2 meters in diameter.

10. How many of these full-grown leaves would fit on the floor of your classroom without overlapping?

Suppose you investigated the growth of a *Victoria amazonica* leaf and drew the following pictures on graph paper, one for every week.

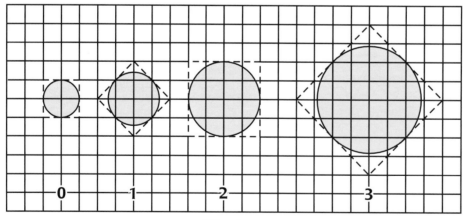

Week

11. Cut out the leaves on **Student Activity Sheet 10.** Paste them onto another sheet of paper so that their centers overlap. How can you tell that the radius of the leaf does not grow linearly?

10. Estimates will vary, depending on the size and layout of your classroom.

11. Answers will vary. Some students may say that they can tell from the differences in sizes of the leaf that the radius does not grow the same amount every week.

Materials Student Activity Sheet 10 (one per student)

Overview Students investigate the growth pattern of the radius of a water lily. They compare the lily's radius at different growth stages to determine whether or not its growth pattern is linear.

About the Mathematics The shape of the water lily leaf used here approximates the general shape of a circle. Therefore, the pictures can be used to represent a growth model in a realistic situation.

Planning Students may work on problems **10** and **11** in pairs or in small groups. Have a class discussion in which students explain the strategies they used to find their answers.

Comments about the Problems

10–11. Although the context of this problem is more difficult than the problems involving bacteria on Student Book page 43, the problems involve the same mathematical concepts.

10. Encourage students to estimate to get an approximate answer for this problem.

11. Note that the overlapping leaves are similar to the rings of a tree. See Section C, "What's Next?" on Student Book pages 31 and 32.

12. Make a table showing the length of the radius of the lily each week. Try to make the numbers as accurate as possible.

13. Make a new table showing the area of the lily each week. To find the area, you can use this formula:

$$\textbf{Area} = \pi r^2$$

14. How can you calculate the area of the NEXT leaf from the area of any CURRENT leaf?

15. What is the growth factor for the area?

16. Does the radius have the same growth factor? If not, what do you think this growth factor might be?

Summary

Many weeds spread rapidly. If you regularly check the area they cover, you may notice that the area increases by a growth factor. Suppose a weed covers 400 square kilometers of area the first year, 800 square kilometers the second year, and 1,600 square kilometers the third year. The table below shows that the area is being measured yearly and that the weed has a growth factor of two. Notice that having a growth factor of two means that the area covered by the weed is doubling every year.

Year	1	2	3	4	5
Area (in km²)	400	800	1,600	3,200	6,400

The water lilies and bacteria shown in this section all had growth factors of two.

12. Tables will vary. Sample table:

Week	0	1	2	3
Radius (mm)	5	7	10	14

13. Tables will vary. Sample table:

Week	0	1	2	3
Area (mm²)	78.5	153.9	314	615.4

14. Answers will vary. Sample response:

If you double the CURRENT area you will get an estimate of the NEXT area: 2 × CURRENT = NEXT, or NEXT = 2 × CURRENT.

15. The growth factor is about two.

16. No. Explanations will vary. Sample explanation:

If you double the radius from one week, you will not get the radius for the next week. The radius grows by a factor of about 1.4.

Materials centimeter rulers (one per student); calculators (one per student)

Overview Students investigate the growth factor for the radius and area of the water lily leaf. They then read and discuss the Summary.

About the Mathematics The growth pattern of the leaf's area can be investigated several ways. One method involves calculating the leaf's area each week using the formula $3.14 \times r \times r$. A second method involves using the approximate area of a circle formula, $3 \times r \times r$ (as used in problem **12** on Student Book page 32). Most students will recognize the doubling pattern in this growth situation and identify the growth factor (two) correctly.

Planning Read and discuss the main mathematical concepts in the Summary with students after they complete problems **12–16**. Ask students how they can see from the numbers in the table in the Summary that the growth factor is two.

Comments about the Problems

12. You may want to point out that students can measure the drawings of the water lily leaf on Student Book page 44 because it shows the actual leaf size.

13. Some students may want to simply add one more row to their existing radius table. Encourage them to make a new table here; the second table will be helpful in solving problem **14.**

14. Since the differences do not show a regular pattern, students may realize that the growth pattern is not linear but exponential, and a growth factor must be found. Most students will see that the growth factor in this pattern is two. Allow students to express this relationship in words. Discuss how they could get a formula from their verbal description.

16. Encourage students to devise their own strategies and use their calculators to find the growth factor. By this point, most students may understand that they must find a number that can be multiplied by the length for week 0 to get the length of week 1. Once that number is found, they can investigate whether the length for week 1 multiplied by that number will equal the length for week 2, and so on. Do not give students an algorithm for finding the growth factor.

Summary Questions

17. Make a table that shows a growth factor of three. Make up the situation and the numbers for your table.

18. The aquatic weed *Salvinia auriculata* (see page 41) also spreads by an annual growth factor. This factor is not two, but another number. Use your calculator to find this decimal growth factor.

Suppose that you get a wonderful job for half a year and that you are allowed to choose how you will be paid:

$1,000 every week	or	1 cent the first week,
		2 cents the second week,
		4 cents the third week,
		8 cents the fourth week,
		and so on . . .

19. Which would you choose? Why?

17. Tables will vary. Sample table:

Time	0	1	2	3	4
Number of Rabbits	2	6	18	54	162

× 3   × 3   × 3   × 3

18. The growth factor is about 1.5. Strategies will vary. Some students may use a table. Sample table:

Year	1959	1960	1961	1962	1963
Area (km²)	200	300	450	675	1012.5

× 1.5   × 1.5   × 1.5   × 1.5

19. Answers and explanations will vary. Sample response:

Doubling a penny adds up to more money in half a year than receiving $1,000 a week. I calculated how much money I would make after half a year using the first case:

26 weeks × $1,000 per week = $26,000.

For the second case, I calculated the amount I would receive week by week with a calculator:

In week ten, the amount would be $5.12.

In week twenty, the amount I would make would already be $5,242.88.

After 23 weeks, my pay for one week would already be $41,943.04, so for one week I would make more than I would get in half a year in the first case.

Overview Students solve three problems about situations of exponential growth.

Planning Students may work on problems **17–19** individually. You might check to see how well each individual student understands exponential growth. Problems **17** and **18** may be used as informal assessment. After students complete Section D, you may assign appropriate activities from the Try This! section, located on pages 54–57 of the *Ups and Downs* Student Book, for homework.

Comments about the Problems

17. Informal Assessment This problem assesses students' ability to understand and use the concept of growth factor and to use algebraic models to represent realistic situations.

You may want to discuss the recursive formula that describes this growth situation: CURRENT × 3 = NEXT.

18. Informal Assessment This problem assesses students' ability to understand and use the concept of a growth factor, and to identify a growth factor. This problem also assesses students' ability to use algebraic models to represent realistic situations.

You might encourage students to use a trial-and-error strategy with their calculators. If students are having difficulty, you might give the hint that they are allowed to round off the area of 1959 from 199 to 200.

Writing Opportunity Problem **19,** which many students find challenging, may be assigned as homework. You might ask students to write in their notebooks about which choice they would make and why.

Work Students Do

Students begin this section by investigating the price of a car that decreases by 50 percent every two years. This pattern of decrease is also examined by the students in the context of medicine absorption in humans. These situations help students develop their understanding of what happens when amounts decrease by a factor of $\frac{1}{2}$. Students end this section by learning about the "half-life of carbon 14" and Geiger counters. They use carbon 14 dating to calculate the age of a ground sloth's skull, of a tree, and of a saber-toothed tiger fossil.

Goals

Students will:

- identify and describe patterns of increase and/or decrease from a table or graph;
- understand and use the concept of growth factor;
- reason about situations of growth in terms of slope, maximum and minimum, range, decrease, and increase;
- use algebraic models to represent realistic situations;
- make connections between situation, graph, and table;*
- recognize the power of graphs and/or tables for representing and solving problems.*

  * These goals are addressed in this section and assessed in other sections of the unit.

Pacing

- approximately three 45-minute class sessions

Vocabulary

- half-life

About the Mathematics

In this section, patterns of decrease known as *exponential decay* are informally introduced. Like exponential growth, exponential decay is essentially repeated multiplication. Exponential decay is an exponential function in which the growth factor is a number between 0 and 1. Within the contexts in this section, a growth factor of one-half, or 50%, is used, since it is the simplest type of exponential decay. This pattern of decrease, with one-half of the previous amount remaining after each time period, is first examined by students in the context of the falling value of a car. For example, the value of a car that is four years old and loses one-half of its value every two years can be calculated using repeated multiplication: $\$10,000 \times 0.5 \times 0.5 = \$2,500$.

Exponential decay is also examined within the context of the amount of medicine absorbed by the body after different time periods. For example, if the initial amount of medicine taken is 500 milligrams and one-half of the medicine is absorbed by the body every hour, then the amount of medicine remaining after three hours can be determined by repeated multiplication: $500 \times 0.5 \times 0.5 \times 0.5 = 62.5$ milligrams left.

Students are also introduced to the term *half-life* in reference to carbon 14 dating. See page 50 of the Student Book for a complete explanation of carbon 14 dating and half-life.

Materials

- graph paper, pages 107 and 111 of the Teacher Guide (two sheets per student)
- calculators, pages 115 and 117 of the Teacher Guide (one per student)
- millimeter graph paper, page 117 of the Teacher Guide (one sheet per student)

Planning Instruction

You may introduce the first context in this section with a brief discussion about how most cars depreciate, or lose their value, over the life of the car. Although most cars do not lose one-half of their value every two years, as used in the context here, you might mention that many cars lose about 10–13% of their value every year. You could also talk about how some cars actually increase their value over the life of the car and become collector's items. You may want to review the concept of percents and how to compute with percents before students begin working through the problems in this section.

Because the problems in this section are rather difficult, you may want students to work through all of the problems in pairs or in small groups. You may also decide to have them work individually on the informal assessment problems 10 and 17–19.

Problems 12–16 are optional. If time is a concern, you may omit these problems or assign them as homework.

Homework

Again, because the problems in this section are rather difficult, you may want students to complete all problems in class rather than as homework assignments. If your students understand the concepts involved and are able to solve the problems with little difficulty, you could assign problems 8–10 (page 110 of the Teacher Guide) as homework. The Writing Opportunities (pages 107 and 117 of the Teacher Guide) also may be assigned as homework. After students complete this section, you may assign appropriate activities from the Try This! section, located on pages 54–57 of the *Ups and Downs* Student Book. The Try This! activities reinforce the key mathematical concepts introduced in this section.

Planning Assessment

- Problem 10 may be used to informally assess students' ability to understand and use the concept of growth factor.
- Problems 17 and 18 may be used to informally assess students' ability to use algebraic models to represent realistic situations and to reason about situations of growth in terms of slope, maximum and minimum, range, decrease, and increase.
- Problem 19 may be used to informally assess students' ability to identify and describe patterns of increase and/or decrease from a table or graph.

FIFTY PERCENT OFF

Monica is shopping for a new or used car. She compares the prices and ages of midsize cars. She notices that adding two years to the age of a car lowers the price by 50%.

1. Copy the diagram on the right. Graph the value of a $10,000 car over a six-year period.

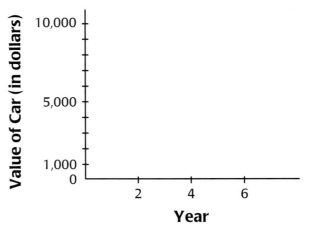

2. Is the graph linear? Why or why not?

3. Monica decides she does not want to keep a car for more than two years. She needs advice on whether to buy a new or used car. Write a few sentences explaining what you would recommend. Support your recommendation.

4. **a.** Show that at this rate, the car never has a zero value.

   b. Is this realistic?

1.

Car Depreciation

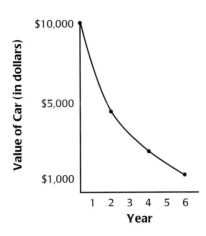

Materials graph paper (one sheet per student)

Overview Students draw a graph showing the decrease in the value of a car over a six-year period. They also decide whether to buy a new or a used car.

About the Mathematics In Section D, students investigated situations that involved exponential growth. Each new value was determined by doubling the original value. In this section, students explore patterns of decrease, known as exponential decay, in which the new value is determined by repeatedly halving the original value. The relationship between percents and fractions is stressed in units from the algebra and number strands.

2. No, the graph is not linear. Explanations will vary. Sample explanation:

The graph is not a straight line because the car's value does not decrease by the same amount every year. Instead, it decreases by a smaller amount each year.

3. Answers will vary. Sample response:

Buy a two-year-old used car. The car is almost new, so it will probably look good and run well. Also, the car has lost its biggest drop in value, so it won't go down in value as much in the years ahead.

4. a. Strategies will vary. Some students may say that extending the graph shows that the curve never reaches the horizontal axis. Other students may say that repeated halving never results in a product of zero, and may use an arrow string to show that even fractional values will never reach zero.

$$1 \xrightarrow{\div 2} \tfrac{1}{2} \xrightarrow{\div 2} \tfrac{1}{4} \xrightarrow{\div 2} \tfrac{1}{8} \xrightarrow{\div 2} \tfrac{1}{16}$$
$$\xrightarrow{\div 2} \tfrac{1}{32} \xrightarrow{\div 2} \tfrac{1}{64} \xrightarrow{\div 2} \tfrac{1}{128} \xrightarrow{\div 2} \tfrac{1}{256}$$

b. Answers will vary. Sample responses:

No, it is not realistic. Some cars may become involved in an accident and be so damaged that you cannot drive them, so these cars have a zero value.

Yes, it is realistic. Some cars may become collector's items and never reach a value of zero. In fact, their value may increase instead of decrease.

Planning You might begin with a class discussion of the context of problem **1.** Ask students to explain what it means that the car price lowers by 50%. Students may work in pairs or small groups on problems **1–4.** Be sure to discuss students' solutions and explanations.

Comments about the Problems

1. Be sure students understand the basic concept of a repeated percent decrease, or exponential decay. They should be able to use the relationship between percents and fractions: lowering a price by 50% is the same as taking half of the price.

3. Encourage students to refer to the pattern of decrease in their arguments. They may give examples in dollar amounts. Students should be able to see that the greatest decrease in price occurs during the first two years of a car's life. After that time, the amount of depreciation is less.

Writing Opportunity You might have students write a few paragraphs in their journals that explain the pros and cons of buying a new car and of buying a used car.

Medicine

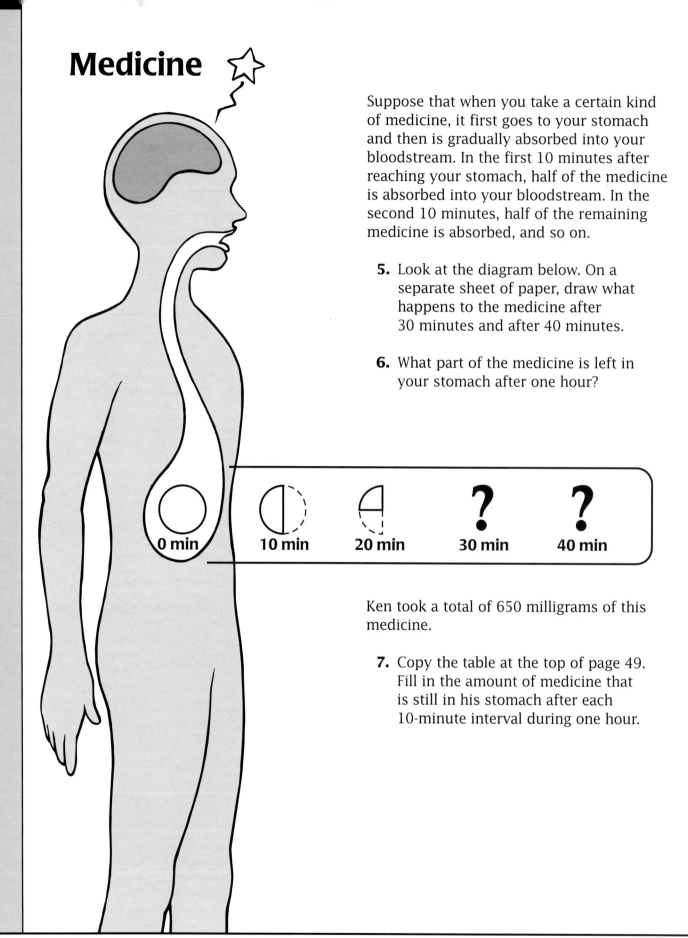

Suppose that when you take a certain kind of medicine, it first goes to your stomach and then is gradually absorbed into your bloodstream. In the first 10 minutes after reaching your stomach, half of the medicine is absorbed into your bloodstream. In the second 10 minutes, half of the remaining medicine is absorbed, and so on.

5. Look at the diagram below. On a separate sheet of paper, draw what happens to the medicine after 30 minutes and after 40 minutes.

6. What part of the medicine is left in your stomach after one hour?

0 min   10 min   20 min   30 min   40 min

Ken took a total of 650 milligrams of this medicine.

7. Copy the table at the top of page 49. Fill in the amount of medicine that is still in his stomach after each 10-minute interval during one hour.

5.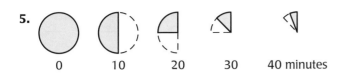

0   10   20   30   40 minutes

6. After 60 minutes, there is $\frac{1}{64}$ of the medicine present.

Some students might show this with a drawing. Sample drawing:

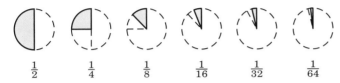

$\frac{1}{2}$ $\frac{1}{4}$ $\frac{1}{8}$ $\frac{1}{16}$ $\frac{1}{32}$ $\frac{1}{64}$

7.

Minutes after Taking Medicine	Medicine in Ken's Stomach (in milligrams)
0	650
10	325
20	162.5
30	81.25
40	40.625
50	20.313
60	10.156

Overview Students investigate the way medicine is absorbed by the bloodstream over time. They make a table of the amount of medicine that remains in a person's stomach after specific time intervals.

About the Mathematics It is critical that students understand and are able to use the relationships between fractions, decimals, and percents to investigate exponential decay in this section. These connections are studied in units such as *Per Sense*, *Fraction Times*, and *More or Less*. For example, to calculate one-half of the medicine amount left after 10 minutes, students may either multiply the amount by $\frac{1}{2}$, find 50% of the amount, or divide the amount by 2, since these expressions are equivalent to each other.

Many students may be able to recognize the similarities between exponential growth, studied in the previous section, and exponential decay. Both situations involve repeated multiplication using a growth or decay factor. In exponential growth, the factor is a number greater than one, while in exponential decay, the factor is a number between zero and one. Students may recall the factors of enlargement or reduction from the unit *Ratios and Rates*, and remember that a factor between zero and one indicates a reduction.

Planning Students may work on problems **5–7** in pairs or in small groups. Be sure to discuss students' solutions and strategies, especially for problem **7.**

Comments about the Problems

6. Encourage students to extend the diagram they drew for problem **5** to help them answer this question. Suggest that they also include a short explanation that describes how they found their answer.

7. Some students may find the answer by using the fractions from problem **5**. For example, after 30 minutes you have $\frac{1}{8}$ left and $\frac{1}{8}$ of 650 is 81.25. Some students may prefer to do repeated division by two. You may want students to round off the amounts to whole numbers or to the nearest tenth.

Did You Know? Some students may ask whether medicine stays in the body for the rest of a person's life. The answer is, of course, no. The halving concept used here is an approximate model that is applicable for a limited time period. Most of the medicine will absorb in the bloodstream; some of it may never be absorbed and will later be expelled from the body.

Minutes after Taking Medicine	Medicine in Ken's Stomach (in milligrams)
0	650
10	
20	
30	
40	
50	
60	

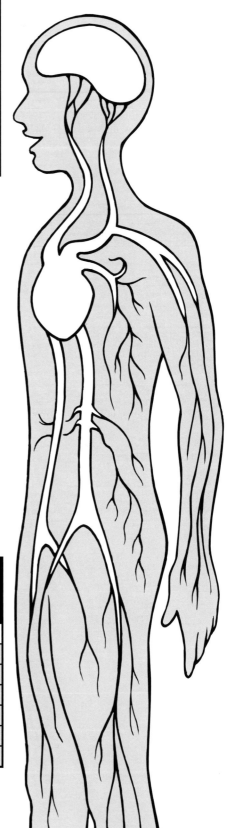

8. Graph the information in the table you just completed. Describe the shape of the graph.

9. Is the amount of medicine in Ken's stomach consistent with your answer to problem **6?** Explain.

Suppose that Ken takes 840 milligrams of another type of medicine. For this medicine, half of the amount in his stomach is absorbed into his bloodstream every two hours.

10. Copy and fill in the table below to show the amounts of medicine in Ken's stomach.

Hours after Taking Medicine	Medicine in Ken's Stomach (in milligrams)
0	840
2	
4	
6	
8	
10	
12	

8. See the graph below. Descriptions will vary. Sample descriptions:

The graph is falling less and less. That is because the amount decreases less and less.

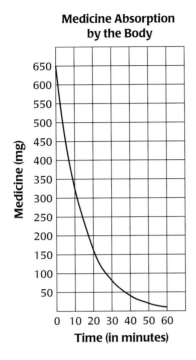

Medicine Absorption by the Body

9. Students should say yes, because $\frac{1}{64}$ of 650 milligrams is about 10 milligrams.

10.

Hours after Taking Medicine	Medicine in Ken's Stomach (in milligrams)
0	840
2	420
4	210
6	105
8	52.5
10	26.25
12	13.125

Materials graph paper (one sheet per student)

Overview Students use the completed table to make a graph. They then investigate a similar problem about the absorption of medicine into a person's bloodstream.

Planning Students may work on problems **8** and **9** in pairs or in small groups. After discussing their answers, you might want them to work on problem **10** individually. Problem **10** may be used as informal assessment. All three problems may be assigned as homework.

Comments about the Problems

8–10. Homework These problems may be assigned as homework.

8. You may check students' graphs while they are working. It is important that students understand that the next amount can be found by taking half of the current amount or by multiplying the current amount by one-half. This relationship can be shown in the table by using arrows. You might want to discuss with students that the amount decreases by a *factor* of one-half, or 0.5.

10. Informal Assessment This problem assesses students' ability to understand and use the concept of growth factor.

Carbon Dating

In Section A, you saw that you could date trees by counting their rings. Another way to date things that were once alive is with *carbon 14 dating.* Carbon 14 is a radioactive type of carbon that is commonly found in Earth's atmosphere. Like all radioactive substances, carbon 14 continuously emits tiny, invisible particles in a process called *radiation.*

All living things have about the same level of carbon 14 because they take in carbon in the form of carbon dioxide. After a plant or an animal dies, the carbon 14 that was present in it decays. The amount of radiation given off can be measured with a Geiger counter.

Carbon 14 decays at a particular rate. Its rate of decay is described in terms of its *half-life,* or the time it takes for one-half of a given amount of carbon 14 to decay.

Scientists have found that the half-life of carbon 14 is 5,730 years; it will lose half of its radioactivity in that amount of time.

Suppose this ground sloth's skull had 16 million atoms of carbon 14 in it when the sloth died many centuries ago. After 5,730 years—the half-life of carbon 14—the skull would have half that many carbon 14 atoms, or 8 million. In another 5,730 years, the number of carbon 14 atoms would be halved again.

11. How many carbon 14 atoms would the skull have after 11,460 years? 17,190 years?

This method of dating is useful only for items that are younger than 30,000 years old, because eventually there is too little carbon 14 left to be measured.

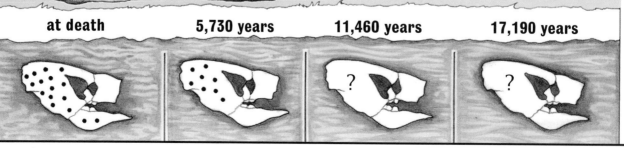

| at death | 5,730 years | 11,460 years | 17,190 years |

11. The skull would have four million carbon atoms after 11,460 years, and two million after 17,190 years.

Strategies will vary. Sample strategy using a ratio table:

Time (in years)	0	5,730	11,460	17,190
Number of Atoms (in millions)	16	8	4	2

Overview Students read about the carbon 14 dating method that archaeologists use to determine the age of fossils. Students then determine the number of carbon 14 atoms in a fossil after a certain number of years.

Planning You might want to read and discuss the text on Student Book page 50 with students. Be sure that they understand the meaning of the term *half-life*: the time it takes for one-half of a given amount of carbon 14 to decay. You can use the context of problem **5** as an example, saying that the half-life of the medicine was 10 minutes. Students may work on problem **11** in small groups.

Comments about the Problems

11. Although this context may be difficult for some students to understand, they may find this topic very interesting. You might suggest that students organize their information in a table (see the solutions column).

Did You Know? Relatively speaking, the amount of radiation given off from carbon 14 is minimal compared to the radiation given off by other elements in the atmosphere.

A Geiger counter registers 15.3 counts per minute for a living tree. From her knowledge of the half-life of carbon 14, one scientist calculated that 1,000 years after the tree died, 88.6% of the carbon 14 would still remain.

The scientist wrote a formula for this: NEXT = CURRENT × 0.886

(time period: 1,000 years)

12. a. Copy and complete the following table.

Years Dead	0	1,000	2,000	3,000	4,000	5,000	6,000
Number of Counts per Minute	15.3						

× 0.886

b. Note that after about 5,730 years, the count is half of the original 15.3 counts per minute. Why does this make sense?

A Geiger counter registers the same number of counts per minute for fossils that are the same age. Therefore, the table applies to all fossils, not just this tree.

12. a.

Years Dead	0	1,000	2,000	3,000	4,000	5,000	6,000
Number of Counts per Minute	15.3	13.56	12.01	10.64	9.43	8.35	7.4

× 0.886

b. Explanations will vary. Sample explanation:

After 5,730 years there is only one-half as much carbon 14 present, so it makes sense that the Geiger counter registers one-half the original number of counts.

Materials calculators (one per student)

Overview Students use a formula to calculate the number of counts per minute that a Geiger counter measures in relation to the amount of carbon 14 in fossils.

About the Mathematics On Student Book page 50, students learned the rate of decay for carbon 14. They learned that after 5,730 years, one-half of the original amount of carbon 14 will remain. On Student Book page 51, students work with a formula that shows how much carbon 14 remains after 1,000 years. For each time period of 1,000 years, 88.6%, or 0.886, of the original carbon amount will remain. When the time period is changed, the factor must also be changed to describe the same situation of decay. The table below shows that one-half of the carbon amount remains between the ages of 5,000 and 6,000 years.

Time (in years)	0	1,000	2,000	3,000	4,000	5,000	6,000
Number of Atoms (in millions)	16	14.2	12.6	11.1	9.9	8.7	7.7

Planning Discuss how Geiger counters work. Some students may have heard of this instrument, but most will not know what it measures and how it is used. Students should understand that the number of counts is a measurement of the amount of carbon 14. So, 15.3 counts per minute represents a certain amount of carbon 14. Students may work on problem **12** in small groups. This is an optional problem. If time is a concern, you may omit this problem or assign it as homework.

Interdisciplinary Connection You might ask a science teacher in your school to possibly bring in a Geiger counter and some fossil samples as a way to introduce the context on Student Book page 52: dating fossils.

A Geiger counter registers 2.34 counts per minute for this saber-toothed tiger fossil.

13. Expand the table you made for problem **12a** to find the age of this fossil.

14. a. Use the data in your table to make a graph.

b. Use your graph to find the age of the fossil.

c. Is it easier to use the table or the graph to find the age of the fossil?

The dead pine tree on the left, a *Pinus aristata*, was found in the White Mountains of California. In Section A, you learned that by counting the annual rings, you can estimate the tree's age.

Suppose you counted the rings and estimated that the tree is about 2,500 years old. A scientist wants to verify this age estimate with carbon dating. A Geiger counter registers 10.1 counts per minute.

15. What is the estimated age of the tree using carbon 14 dating?

16. Which method would you use to estimate the age of the tree—carbon 14 dating or counting rings? Why?

13. The saber-toothed tiger fossil is between 15,000 and 16,000 years old.

0	1,000	2,000	3,000	4,000	5,000
15.3	13.56	12.01	10.64	9.43	8.35

6,000	7,000	8,000	9,000	10,000	11,000
7.40	6.56	5.81	5.15	4.56	4.04

12,000	13,000	14,000	15,000	16,000
3.58	3.17	2.81	2.49	2.21

14. a.

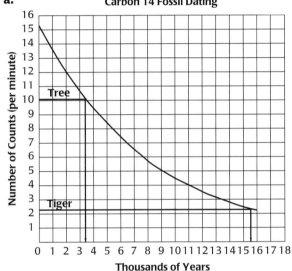

Carbon 14 Fossil Dating

Number of Counts (per minute) vs. *Thousands of Years*

b. Estimates will vary. The fossil is about 15,500 years old. Accept estimates in the range of 15,300–15,700 years old.

c. Answers will vary. Some students may find the graph easier to use because it gives more information and is more precise for numbers that are not listed in the table. Other students may favor the table, saying that drawing the graph is difficult or time consuming.

15. Estimates will vary. The fossil is about 3,400 years old. Accept estimates in the range of 3,200–3,600 years old. (See the graph for problem **14a.**)

16. Answers and explanations will vary. Sample responses:

I would count the rings because this method might be more precise.

I would use the Geiger counter because it is easier to use, especially if there are too many rings to count.

Materials millimeter graph paper (one sheet per student); calculators (one per student)

Overview Students use information from a given table and construct a graph to estimate the ages of two different fossils.

Planning Students may work in pairs or in small groups on problems **13–16.** Be sure to discuss students' answers and strategies with the whole class. These problems are optional. If time is a concern, you may omit these problems or assign them as homework.

Comments about the Problems

13. If students are having difficulty, you might suggest that they round off the numbers in the table to one decimal place. You might also introduce the constant key feature on a calculator.

14. c. Students may argue that the table is easier to use because 2.34 is almost exactly halfway between 2.49 and 2.21. Note that this strategy is precise only when there is a linear relationship and the graph is a straight line. Since the graph is almost a straight line between 15,000 and 16,000 years, this estimation strategy can be used here.

16. Be sure that students realize that dating by using a tree's annual rings can be carried out only if scientists have enough information about that type of tree in that particular region to make a diagram showing its approximate age. The local region is important because the growth of the rings depends on the climatic circumstances. Carbon dating is not dependent on variables such as the tree's location.

Writing Opportunity You might have students write their explanations for problem **16** in their journals.

Summary

You have seen several situations in which amounts have decreased by a factor of $\frac{1}{2}$. For example, the price of a car decreased by a factor of $\frac{1}{2}$ every two years. The amount of medicine in the stomach decreased by a factor of $\frac{1}{2}$ every 10 minutes or every two hours.

It takes 5,730 years for a given amount of carbon 14 to decrease by a factor of $\frac{1}{2}$. The time it takes for something to reduce by half is called its half-life. The half-life of carbon 14 can be used to estimate the ages of such things as fossils and trees.

Summary Questions

Suppose you have two substances, A and B, that are changing in the following ways over the same length of time.

Substance A: NEXT = CURRENT $\times \frac{1}{2}$

Substance B: NEXT = CURRENT $\times \frac{1}{3}$

17. Are the substances increasing or decreasing over time?

18. Which of the substances is changing more rapidly? Support your answer with a table or a graph.

19. Although one of the substances is changing more rapidly, both are decreasing in a similar way. How would you describe the way they are changing — faster and faster, linearly, or slower and slower? Explain.

17. Decreasing. Explanations will vary.
Sample explanations:

For substance A, after every time interval, half of what there was is left; and for substance B, after every time interval, one-third of what there was is left.

I can see the decrease when I use the formulas to make tables:

Time	0	1	2
A	30	15	7.5

$\times \frac{1}{2}$ $\times \frac{1}{2}$

Time	0	1	2
B	30	10	3.3

$\times \frac{1}{3}$ $\times \frac{1}{3}$

18. Substance B is decreasing more rapidly. Tables will vary. See the answer to problem **17** above for the sample tables.

Sample graph:

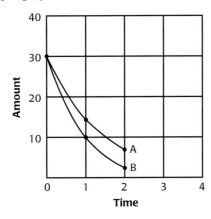

19. Slower and slower. Explanations will vary. Sample response:

You can see by looking at the differences in the tables or graphs, for example:

Time	0	1	2
A	30	15	7.5

-15 $-7\frac{1}{2}$

Time	0	1	2
B	30	10	3.3

-20 -7

Overview Students read the Summary and investigate two recursive formulas that describe situations of change over time.

Planning Read and discuss the Summary with students, focusing on the similarities and differences between exponential increase and decrease. In situations involving exponential increase, the factor is a number greater than one, while in situations involving exponential decrease, the factor is a number between zero and one. You might want students to work on problems **17–19** individually. These problems may also be used as informal assessment. After students complete Section E, you may assign appropriate activities from the Try This! section, located on pages 54–57 of the *Ups and Downs* Student Book, for homework.

Comments about the Problems

17–18. Informal Assessment These problems assess students' ability to use algebraic models to represent and investigate realistic situations and to reason about situations of growth in terms of slope, maximum and minimum, range, decrease, and increase.

Note whether students identify and use the growth factor in both situations. If so, this may be an indication that they have understood the concept of exponential decrease.

17. You might encourage students to argue by using a table or a graph with a particular example.

19. Informal Assessment This problem assesses students' ability to identify and describe patterns of increase and/or decrease from a table or graph.

Assessment Overview

Students work on eight assessment problems that you can use to collect information about what each student knows and understands about different situations of growth and what strategies they used to solve the problems.

Goals

- use information about increase and/or decrease to create line graphs
- identify and describe patterns of increase and/or decrease from a table or graph

- identify characteristics of periodic graphs

- identify linear patterns in tables and graphs

- understand and use the concept of growth factor

- describe linear growth with recursive formulas
- describe linear growth with direct formulas

- make connections between situation, graph, and table

- reason about situations of growth in terms of slope, maximum and minimum, range, decrease, and increase

- identify a growth factor

- recognize the power of graphs and/or tables for representing and solving problems

- use algebraic models to represent realistic situations

Assessment Opportunities

Dots, Distances, and Speed
Taxi!

Make Up a Story
Dots, Distances, and Speed
Taxi!
Bacteria in Food
Radioactivity

Deep Water
Lighthouse

Taxi!
Growth

Growth
Bacteria in Food
Radioactivity

Growth

Taxi!
Growth

Make Up a Story
Dots, Distances, and Speed

Make Up a Story
Dots, Distances, and Speed
Bacteria in Food

Bacteria in Food
Radioactivity

Deep Water
Taxi!
Bacteria in Food

Lighthouse
Taxi!
Radioactivity

Pacing

- When combined, the eight assessment activities will take approximately two 45-minute class sessions. See the Planning Assessment section for further suggestions as to how you might use the assessment activities.

About the Mathematics

These eight end-of-unit assessment activities evaluate the major goals of the *Ups and Downs* unit. Refer to the Goals and Assessment Opportunities sections on the previous page for information regarding the goals that are assessed in each assessment activity. Students may use different strategies to solve each problem. Their choice of strategies may indicate their level of comprehension of the problem. Consider how well students' strategies address the problem, as well as how successful students are at applying their strategies in the problem-solving process.

Some activities based on exponential growth, such as the Radioactivity assessment, are rather difficult and may be challenging for some students. At this point, students should demonstrate an understanding of linear and exponential growth patterns rather than a mastery of these mathematical concepts. For example, many students may continue to have difficulty identifying the exponential growth factor or writing a formula to describe these growth patterns. The concepts of exponential growth and recursive and direct formulas are revisited and further developed in the unit *Growth*.

Materials

- Assessments, pages 153–161 of the Teacher Guide (one of each per student)
- graph paper, pages 129, 135, 137, and 139 of the Teacher Guide (four sheets per student)
- calculators, page 135 of the Teacher Guide (one per student)

Planning Assessment

You may want students to work on these assessments individually, in pairs, or in small groups, depending on the nature of your class and your goals for assessment. It is important, however, that each student participates and formulates his or her own written responses so that you can evaluate each student's understanding and abilities.

Make sure that you allow enough time for students to complete the assessment activities. Students are free to solve each problem in their own ways. They may choose to use any of the models introduced and developed in this unit to solve problems that do not call for a specific model.

Scoring

Answers are provided for all assessment problems. The method of scoring the problems depends on the types of questions in each assessment. Most questions require students to explain their reasoning or justify their answers. For these questions, the reasoning used by the students in solving the problems as well as the correctness of the answers should be considered as part of your grading scheme. A holistic scoring scheme can be used to evaluate an entire task. For example, after reviewing a student's work, you may assign a key word such as *emerging, developing, accomplishing,* or *exceeding* to describe their mathematical problem-solving, reasoning, and communication.

On other tasks, it may be more appropriate to assign point values for each response. Students' progress toward the goals of the unit should also be considered. Descriptive statements that include details of a student's solution to an assessment activity can be recorded. These statements could provide insight into a student's progress toward a specific goal of the unit. Descriptive statements are often more informative than recording only a score and can be used to document student growth in mathematics over time.

MAKE UP A STORY

Use additional paper as needed.

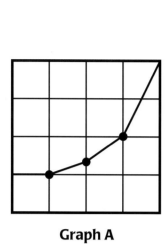 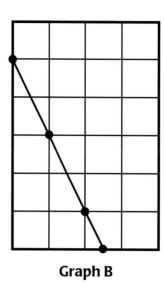 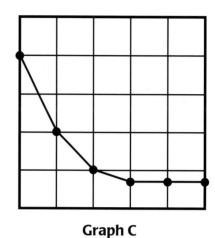

Graph A   **Graph B**   **Graph C**

1. These three graphs each tell a story about a situation in which something increased or decreased. Describe the pattern of increase or decrease for each graph.

2. Choose one graph and copy it on a sheet of paper. Make up a story that goes with that graph. Choose appropriate scales and labels for the horizontal and vertical axes, and include them on your graph.

1. Answers will vary for each graph. Sample responses:

Graph A:
It started as a gradual increase. Then it increased faster and faster.

Graph B:
The straight line shows that the graph decreased at the same rate.

Graph C:
It decreased at a fast rate at first. Then it slowed down and decreased at a slower rate. Then it stayed the same and did not increase or decrease at all.

2. Answers will vary. Sample response:

Temperature Change

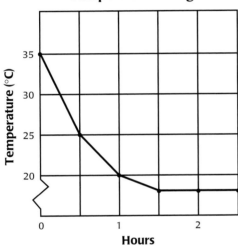

The temperature outside was 35°C as it started to drop. At first, the temperature dropped quickly. Then it dropped at a much slower rate until it was 20°C one hour later. Then the temperature leveled off at 18°C and did not change for one hour.

Materials Make Up a Story assessment, page 153 of the Teacher Guide (one per student)

Overview Students describe the pattern of increase or decrease in three graphs. They also write a story that fits the growth pattern in one graph.

About the Mathematics This activity assesses students' ability to identify and describe patterns of increase and/or decrease from a table or graph and to reason about growth situations in terms of slope, maximum and minimum, range, decrease, and increase. It also assesses students' ability to make connections between situation, graph, and table. These problems are similar to those in Section A of this unit.

Planning You may want students to work individually on these assessment problems.

Comments about the Problems

1. Make sure that students understand that they should include a detailed description here rather than merely indicating that the graph shows increase or decrease. Note whether or not students use the correct wording in their descriptions. For example, phrases such as *It increases fast* or *It increases faster* are incorrect. The correct way to describe the graph would be to say *It increases faster and faster* or *It increases at a faster rate*. Students need not use this exact wording, as long as they describe the situation in an accurate way.

DOTS, DISTANCES, AND SPEED

Use additional paper as needed.

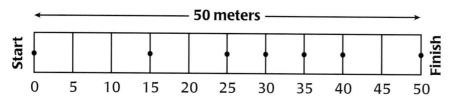

Here you see a strip that shows the movement of a car during a six-second time period. Every dot indicates the place of the car along the road. Between any two dots, the amount of time is one second.

1. What is the total distance the car covered in the first four seconds?

2. Draw a graph that shows how the distance increased over time.

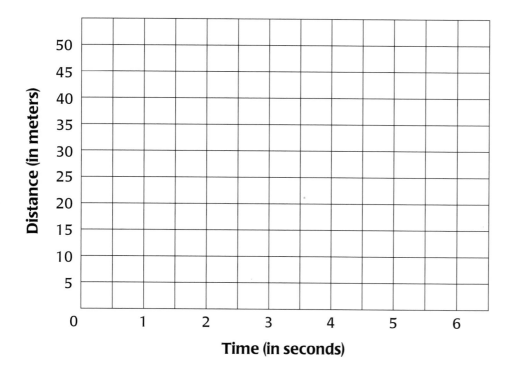

3. When did the car have the fastest speed? Explain.

4. Describe the pattern of increase or decrease in the above graph in your own words.

1. The car traveled a total of 35 meters during the first four seconds.

2.

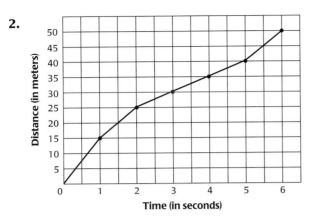

3. The car had the fastest speed during the first second because it covered the greatest distance during that second.

4. Answers will vary. Sample response:

The car reached its fastest speed right after the start. Then it went at a slower rate. After two seconds it maintained a constant speed for three seconds. Then the car increased its speed to the end.

Materials Dots, Distances, and Speed assessment, page 154 of the Teacher Guide (one per student)

Overview Students use the dots on a strip to create a line graph that shows the distance over time. They also describe the speed of the car over time in words.

About the Mathematics These assessment activities essentially evaluate the same goals as the Make Up a Story assessment. It also assesses students' ability to use information about increase and/or decrease to create line graphs. Students again apply the skills and concepts they learned in Section A to a new situation. This activity is similar to the situation on Student Book page 7, which dealt with growth in height and speed of growth. The situation here deals with distance and speed. Students also worked on an activity similar to this in the unit *Tracking Graphs*.

Planning You might want students to work individually on these assessment problems.

Comments about the Problems

1. The intent of this question is to get students to focus on the total distance covered during the entire time period rather than looking at the distance covered by the car during each time period. Students should realize that the first dot represents zero seconds, so the fifth dot represents four seconds.

2. Students may use various methods to graph the given data. Some students may make a table and use the table data to draw a graph. Others may plot the points on the graph directly.

3. Many students may be able to answer this question from the given strip rather than the graph they made, since the information in each location is identical.

4. Allow students to decide how detailed their descriptions of the graph will be. Some students may give a global description (see the solutions column). Other students may give a more detailed answer using numbers. In either case, students' answers should give an indication of their understanding of the relationship between speed and distance in this situation. This is a difficult concept for many students.

DEEP WATER

Use additional paper as needed.

This graph shows how the water depth in a coastal harbor changes over time.

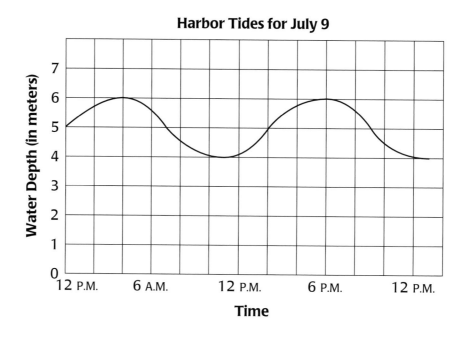

Harbor Tides for July 9

1. Why is this a periodic graph?

2. Color one cycle of the graph.

3. How long is one period of the graph?

4. A captain wants to know at what time on July 10 the water will be at its maximum depth. How can he find this out? What is the answer?

1. This is a periodic graph because it shows a repeating pattern.

2. Answers will vary. Sample response:

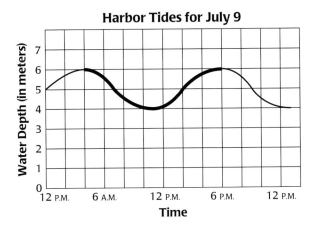

Harbor Tides for July 9

3. One period is about 14 hours.

4. The water will be at its maximum depth at 8:00 A.M. and at 6:00 P.M.

   Strategies will vary. Sample strategies:

   The captain can extend the graph and draw the repeating pattern for the next day.

   The captain can add 14 hours to the last maximum depth time on July 9th. (6:00 P.M. + 14 hours = 8:00 A.M.)

Materials Deep Water assessment, page 155 of the Teacher Guide (one per student)

Overview Students investigate a tidal graph and identify and use the period of the tidal graph to predict at what time the water depth will be at its maximum the following day.

About the Mathematics These assessment activities evaluate students' ability to identify characteristics of periodic graphs. It also assesses students' ability to recognize the power of graphs and/or tables for representing and solving problems. In this assessment, students use the repeating pattern in the tidal graph to make an extrapolation. Students have not used periodic graphs in this way in the unit. Students' answers to problem **4** will give you an indication of how well they understand the characteristics of periodic situations.

Planning You may want students to work individually on these assessment problems.

Comments about the Problems

1–3. Most students will have little difficulty solving these problems.

4. Students may find this to be a challenging problem, since they must decide how to find the solution on their own. They must first understand that the water depth is at its maximum two times per day. Some students may then choose to extend the graph to find these times. Others may use the period of the graph to reason that if the water level is at a maximum height at about 6:00 P.M. on July 9, then it will be at its maximum height again 14 hours later.

LIGHTHOUSE

Use additional paper as needed.

Ships without radar or other modern equipment on board can find their way at night by using light signals from lighthouses. Every lighthouse has its own light signal, so a captain can identify a lighthouse from the pattern of the light flashes.

Below you see a graphical representation of one lighthouse's signal pattern.

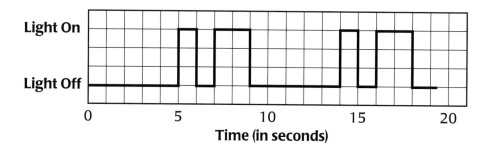

1. Examine the diagram of the lighthouse signal pattern. Shade the diagram to show one cycle of the signal pattern. How long is one period?

2. How many light flashes would you see in a one-minute period?

3. Create your own lighthouse signal pattern. Choose the number of light flashes and the period of the lighthouse signal pattern by yourself. Shade your diagram to show one cycle of your signal pattern.

1.

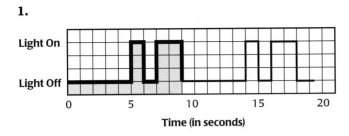

One period is nine seconds.

2. You would see 13 flashes during a one-minute period.

Explanations will vary. Sample explanation:

During the first 54 seconds, you would see six complete cycles of two flashes, or 12 flashes. During the next 6 seconds, you would see one flash of the next cycle. So, altogether, you would see 13 flashes.

3. Patterns will vary. Sample pattern:

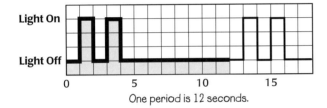

One period is 12 seconds.

Materials Lighthouse assessment, page 156 of the Teacher Guide (one per student); graph paper (one sheet per student)

Overview Students investigate a graphical representation of the light signal pattern of a lighthouse. They determine the period and calculate the number of flashes per minute. Then they create their own light signal pattern.

About the Mathematics This assessment activity evaluates students' ability to identify characteristics of periodic graphs. It also assesses students' ability to use algebraic models to represent realistic situations. It assesses the same goals as the Deep Water assessment but in a new context. The diagram is a schematic graphic representation of the periodic light signal pattern of a lighthouse.

Planning You may want students to work individually on these assessment problems.

Comments about the Problems

1. Most students will be able to correctly identify one period in the graph (nine seconds). Within this context, one nine-second period is also the time span needed for the lighthouse's spotlight to make one complete 360° turn.

2. Students may use various strategies here. Some students may extend the graph to find the total number of flashes. Others may use the time span of one period and its multiples to compute how many whole cycles the pattern can make within one minute (six cycles or 12 flashes in 54 seconds). They would then figure out how many additional flashes would be seen in the remaining six seconds (one flash) and add the number of flashes (13 flashes).

3. Encourage students to show their lighthouse signal pattern in a graphic representation similar to the one in this assessment. Accept verbal descriptions of the patterns that are accurate and complete.

TAXI!

Use additional paper as needed.

In the small town of Bakersfield, there are two taxicab companies. The Metro Taxi Company charges customers a flat rate of $4 per mile. The Acme Taxi Company uses the following table to show their cab fare costs.

Acme Taxi Company Fare Costs

Distance (in miles)	1	2	3	4
Cost (in dollars)	7	9	11	13

1. Draw a graph for *each* taxi company showing the costs of cab rides for various distances.

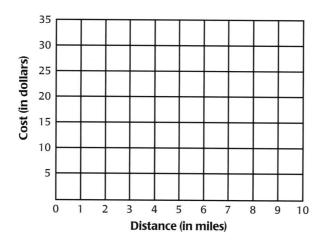

2. Write formulas that could be used to calculate the costs of cab rides using the two taxi companies. Use the letter *C* to represent the cost of the trip and the letter *M* to represent the distance in miles.

3. **a.** Write a proposal to the owner of the Acme Taxi Company, suggesting how she might increase profits by raising taxicab fare rates a little while still staying competitive with the Metro Taxi Company. Explain your plan in detail.

   b. Write a formula that could be used to calculate the cost of a cab ride using your new cab fare rates for the Acme Taxi Company.

1.

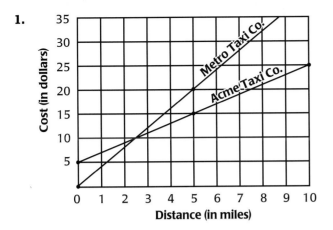

2. Formulas may vary. Sample formulas:

Metro: $C = 4M$ Acme: $C = 5 + 2M$

3. a. Answers will vary. Accept any plan that indicates fare rates lower than or equal to the Metro Taxi Company. Sample plan:

I would suggest that the Acme owner raise the cab ride rates to $5 plus $3 per mile. The owner makes a profit of one dollar more per mile.

    b. Formulas will vary, depending on students' proposed plans in part **a.** Sample formula: $C = 5 + 3M$.

Materials Taxi! assessment, page 157 of the Teacher Guide (one per student)

Overview Students compare the fares of two taxicab companies. They use the given data to make a graph that compares the cab fares of both companies and write formulas that can be used to calculate the costs of cab rides using either company.

About the Mathematics These assessment activities evaluate students' ability to identify linear patterns in tables and graphs and describe linear growth patterns using direct formulas. These problems cover the main goals of Section C.

Planning You might want students to work individually on these assessment activities.

Comments about the Problems

1. Suggest that students label the graph lines to indicate which company each line represents.

2. Students' responses will show their understanding and ability to write formulas to represent growth situations in a given context.

3. Students may find this to be a challenging problem. They have the option of changing the initial charge of $5 or the per mile charge of $2 or both charges. Be sure that students include an explanation for their proposal.

GROWTH

Use additional paper as needed.

The table below can be used to show different types of growth patterns.

Time (in years)	0	1	2	3	4
Length (in mm)	10	20			

1. **a.** Complete the table to show a linear growth pattern.

   b. Write a NEXT-CURRENT formula that describes this situation.

   c. Write a direct formula that describes this situation. Use the letter T to represent time and the letter L to represent length.

2. **a.** Complete the table below to show a growth pattern that increases according to a certain growth factor.

Time (in years)	0	1	2	3	4
Length (in mm)	10	20			

   b. Write a NEXT-CURRENT formula that describes this situation.

3. Describe the shapes of the graphs for both situations: that with the linear pattern and that with the pattern that increases according to a certain growth factor.

Solutions and Samples
of student work

1. a.

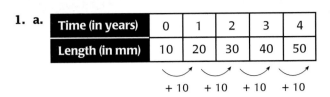

Time (in years)	0	1	2	3	4
Length (in mm)	10	20	30	40	50

+ 10 + 10 + 10 + 10

b. NEXT = CURRENT + 10

c. Formulas will vary. Sample formula:

$L = T \times 10 + 10$

2. a.

Time (in years)	0	1	2	3	4
Length (in mm)	10	20	40	80	160

b. NEXT = CURRENT × 2

3. Answers will vary. Sample response:

The first situation shows linear growth, so the graph is a straight line. The second situation shows growth that increases faster and faster, as you can see from the numbers in the table. So, its graph will get steeper and steeper.

Materials Growth assessment, page 158 of the Teacher Guide (one per student)

Overview Students complete a table of values to show linear growth and complete another table to show growth determined by a growth factor. For both situations, they also write a formula and describe the general shape of its graph.

About the Mathematics These assessment problems evaluate students' ability to identify linear and exponential patterns in tables and graphs; their ability to describe a linear growth pattern with both recursive and direct formulas; and their understanding of the concept of a growth factor. These were the key mathematical goals presented in Sections C and D.

Planning You might want students to work individually on these assessment activities.

Comments about the Problems

1. a. Most students will compare the difference between the first two table values to determine the amount that should be added to make each additional table entry show linear growth. If students are having difficulty, encourage them to look for this difference.

   c. Students may have difficulty writing a direct formula to describe this growth situation. Encourage students to plug values into their formulas to determine whether they are indeed correct.

2. a. Some students may have difficulty making the transition from using addition to make the linear growth pattern and using multiplication to show the exponential growth pattern. Encourage students to look at the given numbers in the table to find the growth factor.

BACTERIA IN FOOD

Use additional paper as needed.

The bacteria salmonella often causes food poisoning. At 35°C, a single bacterium divides every hour. Suppose there are 100 salmonella bacteria in a portion of food, and the temperature is 35°C.

1. Complete the table below to find out after how many hours there will be more than 1,000 bacteria present in the food.

Time (in hours)	0	1	2			
Number of Bacteria	100					

2. On a sheet of graph paper, make a graph that shows the growth pattern of the bacteria in the above table.

3. What is the growth factor for the situation described in problems **1** and **2?**

Suppose that by refrigerating the food, you are able to slow the growth of bacteria to exactly one-half of the growth rate in the previous situation.

4. Make a graph that shows this new growth pattern. Draw the graph on the same coordinate grid system that you used in problem **2.**

5. Estimate the growth factor for the new growth pattern described in problem **4.** Explain how you found your answer.

1. The bacteria population will be over 1,000 between 3 and 4 hours.

Time (in hours)	0	1	2	3	4
Number of Bacteria	100	200	400	800	1,600

2.

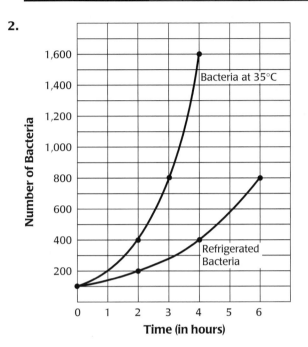

3. The growth factor is 2.

4. See the graph shown in the solution to problem **2.**

   Sample table used to make the graph:

Time (in hours)	0	1	2	3	4	5	6
Number of Bacteria	100		200		400		800

5. The growth factor is about 1.4. Strategies will vary. Sample strategy:

   I used the new graph to fill in the missing values in the new table. Then I looked at the first and second values in the new table to find out what the first number was multiplied by to get the second number. I tried 1.2, 1.3, 1.4, and so on until I found a growth factor that worked.

Time (in hours)	0	1	2	3	4	5	6
Number of Bacteria	100	140	200	280	400	560	800

× ?

Materials Bacteria in Food assessment, page 159 of the Teacher Guide (one per student); graph paper (one sheet per student); calculators (one per student)

Overview Students investigate the population growth of a given number of bacteria and determine the growth factor in this situation.

About the Mathematics These assessment activities evaluate students' ability to understand and identify a growth factor in an exponential growth situation, the main concepts covered in Section D of this unit. Problems **4** and **5** challenge students to analyze and generalize about the concept of growth factor at a higher level than is required for the previous problems.

Planning You may want students to work individually on these assessment problems.

Comments about the Problems

1. Most students will realize that when each single bacterium divides hourly, the total number of bacteria doubles, making the growth factor 2.

4. Encourage students to make a table and fill in values for the number of bacteria in this new situation. See the solutions column for a sample table.

5. Students may use a calculator and look for a growth factor using a trial-and-error method. If students are having difficulty, you might give them the hint that the growth factor is smaller than that in problem **1.** You could also give them different numbers they could try (see the solutions column for a specific strategy).

Use additional paper as needed.

Most nuclear energy is produced by splitting atoms of uranium. This atom-splitting releases great energy that can be used for many purposes. However, the atom may split in different ways to create as many as 200 different products! These products are called *nuclear waste.*

Nuclear waste is highly radioactive, and it is therefore unsafe for living things. Nuclear waste must be kept in a safe place away from people, animals, and food and water supplies until it loses its dangerous level of radioactivity.

Suppose that a nuclear plant produced radioactive waste that would lose its radioactivity as follows: every ten years, the amount of radioactivity decreases by 25%.

One scientist suggested that the nuclear waste be safely stored in barrels for a period of 40 years until it is no longer radioactive. Then the barrels could be dumped into the ocean without causing any harm to humans, marine life, or food and water supplies.

1. How might the scientist have determined that it would take 40 years for the nuclear waste to lose its radioactivity? Do you agree with her plan? Explain your reasoning.

Solutions and Samples
of student work

1. Opinions and explanations will vary. Some students may say that it is incorrect to reason that if 25% of the radiation disappears every 10 years, then all of the radiation will be gone in 40 years (4 × 10 = 40).

Some students may show the radioactive decay using their own starting amount level:

Suppose that the starting amount was 100 counts per minute. In 10 years, you will have 75 counts per minute. In 20 years, you will have 56 counts per minute, and in 40 years, you will have 32 counts per minute.

Materials Radioactivity assessment, page 160 of the Teacher Guide (one per student); graph paper, optional (one sheet per student)

Overview Students read and solve a problem about radioactive waste material.

About the Mathematics These assessment problems evaluate students' ability to understand, identify, and use growth factors and reason about exponential decay. Within the context of the given situation, the radioactivity level decreases by 25% every 10 years. Interpreted correctly, that means that the radioactivity level decreases by 25% of the amount at the start of that 10-year period, not by 25% of the original amount for the entire decaying process being considered. Thus, when it drops 25%, 75% of the current amount will remain. Using the relationship between percents and fractions, you can say that three-fourths of the current amount gives the next amount, so the growth factor in this situation is three-fourths.

Planning You might want students to work in pairs or in small groups on this assessment problem. Students may use graph paper to answer problem **1.**

Comments about the Problems

1. Students may have difficulty reasoning through this problem, since they are not given a specific starting level. The focus here is for students to communicate in their answer that the scientists are not correct. If students can show and verbalize the fact that the decrease is 25% of the remaining amount, they have an understanding of exponential decay.

Use additional paper as needed.

Every 10 years, the amount of radioactivity in the substance decreases by 25%. This gradual decrease can be illustrated by drawing pictures. The picture below represents an initial amount of nuclear waste.

2. Using the above drawing as a model, draw a picture below to show how much of the initial amount of radioactivity is left after 10 years. Label this picture "Ten Years Later." Then draw another picture to show how much of the remaining amount of radioactivity is left after 10 more years. Label this picture "Twenty Years Later."

3. Carefully examine the two pictures you just drew that show the amount of radioactivity left after 10 and 20 years. Look back at your answer to problem **1.** Do you agree with the scientist's plan now? Explain your reasoning.

4. Now imagine a different situation in which a scientist started with nuclear waste material that showed a radioactivity level of 1,024 counts per minute on a Geiger counter. Make a table and a graph on a sheet of graph paper to show how the amount of radioactivity would change over time.

5. Give an estimate of the amount of radioactivity that still remains after 40 years.

6. Compare your answer to problem **5** with your answer to problem **1.** Do you want to revise your answer to problem **1?** Explain.

7. Write a NEXT-CURRENT formula to describe the decreasing amount of radioactivity in this situation.

2.

0 Years

10 Years Later

20 Years Later

3. No, it is not safe. Explanations will vary. Sample explanations:

It takes more than 40 years for all the radioactivity to decay.

Every 10 years, one-fourth of the remaining radioactivity disappears.

40 years later

4. Tables and graphs will vary. Sample response:

Years	0	10	20	30	40
Counts per Minute	1,024	768	576	432	324

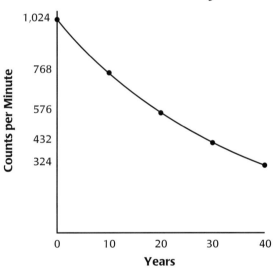

Radioactive Decay

5. Estimates will vary. Sample estimates:

About 30% of the radioactivity remains.
A Geiger counter would register about 324 counts per minute of radioactivity.

6. Answers will vary, depending on students' initial answer for problem **1**.

7. Formulas will vary. Sample formulas:

NEXT = CURRENT − 0.25 × CURRENT
NEXT = CURRENT − $\frac{1}{4}$ CURRENT
NEXT = 0.75 × CURRENT
NEXT = $\frac{3}{4}$ × CURRENT

Materials Radioactivity assessment, page 161 of the Teacher Guide (one per student); graph paper (one sheet per student)

Overview Students continue to solve problems about radioactive waste material.

Planning You might want students to work in pairs or in small groups on this assessment problem.

Comments about the Problems

2–3. These problems ask students to illustrate the given situation with a drawing. Students who did not solve problem **1** correctly may now see that each subsequent box contains 25% less than the previous box.

4–6. More students may now understand the structure of this problem, since they are given a starting amount to work with.

Ups and Downs
Glossary

The Glossary defines all vocabulary words listed on the Section Opener pages. It includes the mathematical terms that may be new to students, as well as words having to do with the contexts introduced in the unit. *(Note:* The Student Book has no glossary. This is in order to allow students to construct their own definitions, based on their personal experiences with the unit activities.)

The definitions below are specific to the use of the terms in this unit. The page numbers given are from this Teacher Guide.

cycle (p. 54) the portion of a periodic graph that repeats itself

growth factor (p. 96) the factor by which an object increases or grows over equal time periods

half-life (p. 112) the time it takes for one-half of an amount to decay

linear growth (p. 70) a growth pattern showing a steady rate of increase; the graph is a straight line

period (p. 54) the amount of time it takes for a periodic graph to repeat itself

periodic (p. 54) repeating the same pattern

Blackline
Masters

Dear Family,

Your child will soon begin the *Mathematics in Context* unit
Ups and Downs. This algebra unit presents real-world contexts that can be represented by graphs that show change over time.

Students investigate graphs of ocean tides, tree growth, temperature changes, blood pressure, radioactive decay, and so on. They gain an understanding of what the shape of a graph means. Why might a graph go only upward? only downward? Students learn to identify cycles (the repeating parts of graphs) and periods (the durations of the parts that repeat).

You can help your child learn about graphs that show change over time by pointing them out when you see them in newspapers or magazines. You may wish to keep track of the temperatures over a week or a month and then create a graph that represents the data you have collected. You can also graph the temperature predictions that your local weather person makes and see how the graphs of the predicted and actual temperatures compare.

We hope you enjoy discussing the uses of graphs and the stories behind them with your child.

Sincerely,

The Mathematics in Context Development Team

Dear Student,

Welcome to *Ups and Downs*. In this unit, you will look at things that change over time, such as blood pressure or the tides of an ocean. You'll learn to represent these changes using tables, graphs, and formulas.

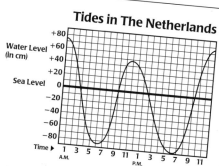

Tides in The Netherlands

Graphs of temperatures and tides go up and down, but some graphs go only upward or only downward, like graphs for tree growth or ice melting.

As you become more familiar with graphs and the changes they represent, you will begin to notice and understand graphs in newspapers, magazines, and advertisements. You will see the advantages of telling a story with a graph.

POPULATION OF SEATTLE GROWING FASTER AND FASTER

PRICE OF CALCULATORS DROPPING FASTER AND FASTER

Sincerely,

The Mathematics in Context Development Team

Name_____

Match core sample C to core samples A and B.

7. What period of time is represented by the three core samples?

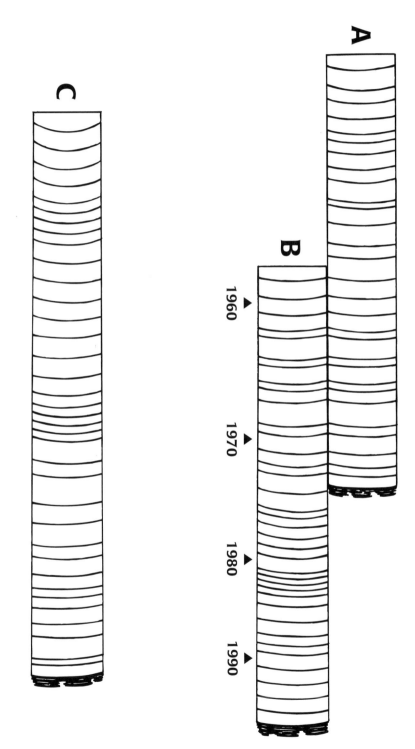

14. Draw a graph of Marsha's growth. The vertical axis has the same scale as the door, so you can use the marks on the door to get the vertical coordinates.

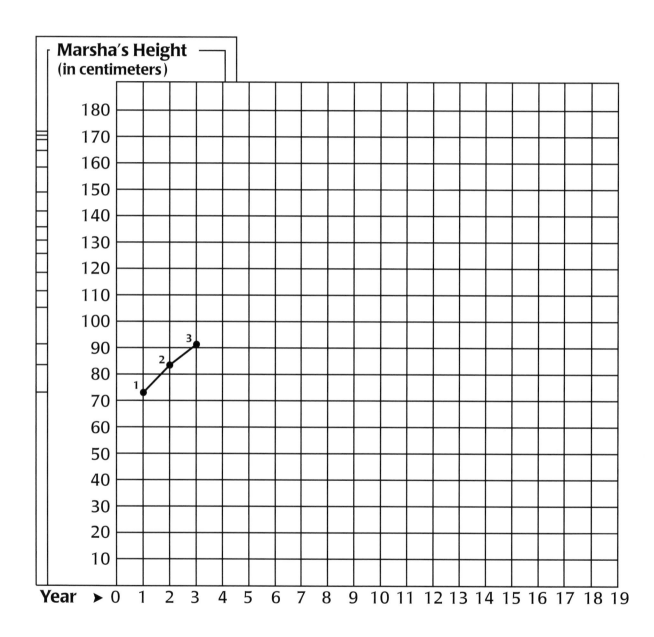

18. Draw a graph of Dean's height.

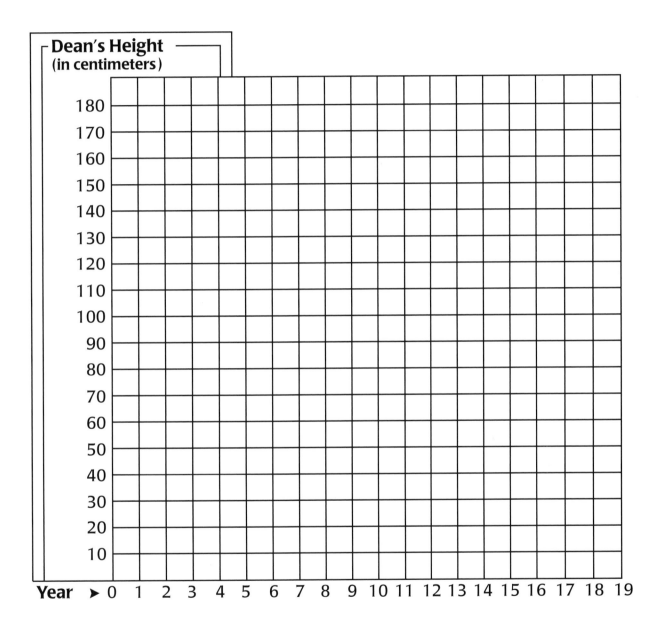

Name _____

WEIGHT GROWTH CHART FOR BOYS
Age: Birth to 36 months

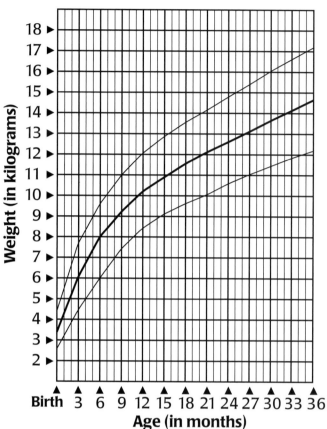

© Am. J. Clin. Nutr.
American Society for Clinical Nutrition

LENGTH GROWTH CHART FOR BOYS
Age: Birth to 36 months

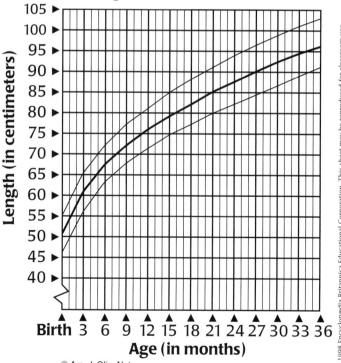

© Am. J. Clin. Nutr.
American Society for Clinical Nutrition

© 1998 Encyclopædia Britannica Educational Corporation. This sheet may be reproduced for classroom use.

5. Use the information in the table on page 23 to sketch a graph of the water levels near the Golden Gate Bridge for these three days.

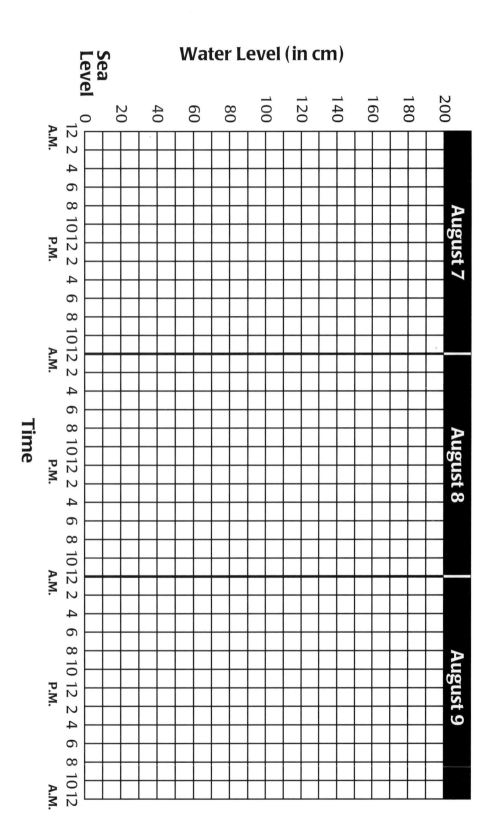

9. What numbers might you write along the horizontal and vertical axes? Label the axes on the graph below with these numbers.

10. b. Color one cycle on the graph below.

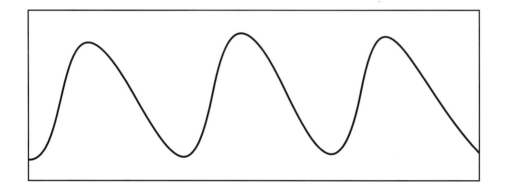

Time (in minutes)

Use the graph below for the following problems.

11. Label the horizontal axis.

12. How long is one period of the graph?

13. Color one cycle on the graph. Describe what is happening to the camel's body temperature during this period.

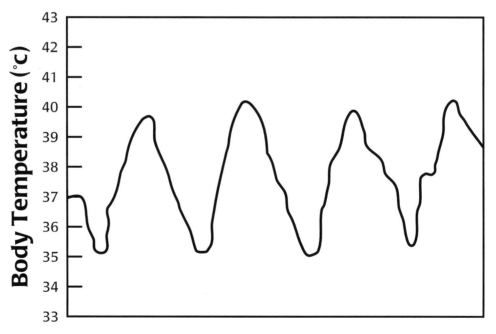

Name_____

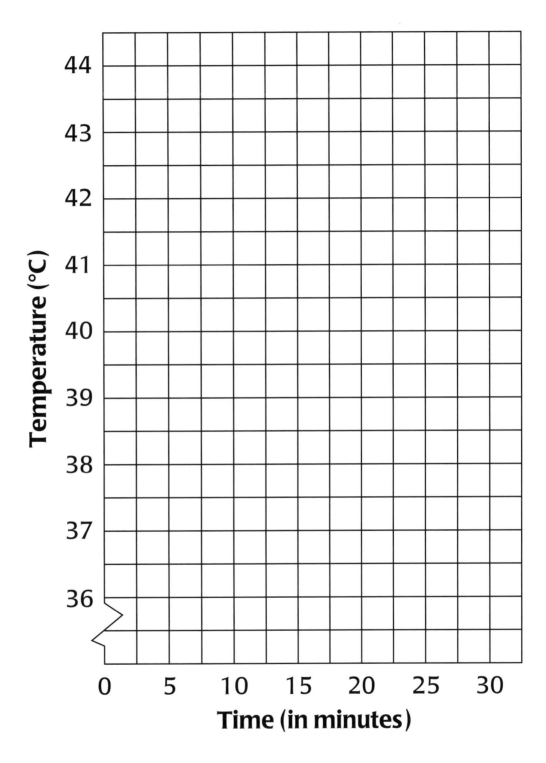

Temperature (°C)

44

43

42

41

40

39

38

37

36

0   5   10   15   20   25   30

Time (in minutes)

Name _____

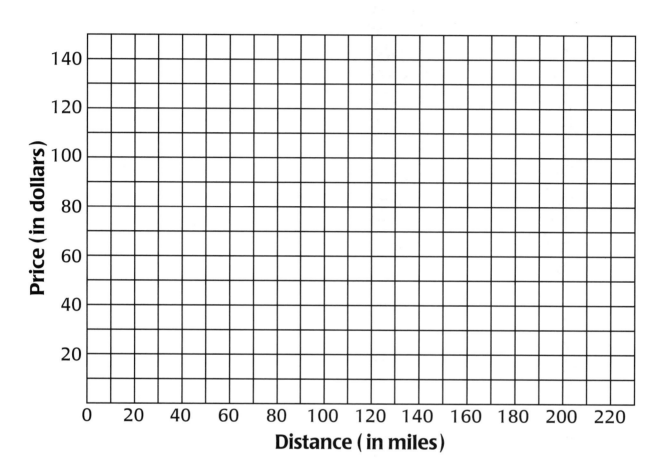

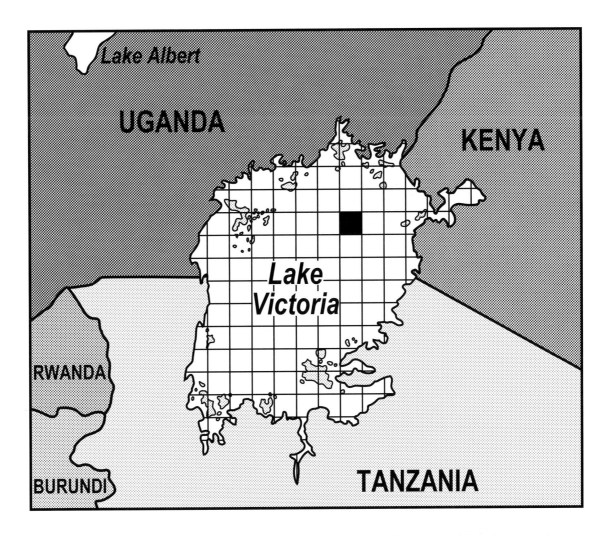

2. If the shaded square represents the area currently covered by the weed, how many squares would represent the area covered next year? a year later? a year after that?

3. Angela is showing the growth of the area covered by the weed by coloring squares on the map. She uses a different color for each year. She remarks: "The number of squares I color for a certain year is exactly the same as the number already colored for all of the years before." Use **Student Activity Sheet 9** to show why Angela is or is not correct.

4. How many years would it take for the lake to be about half covered?

5. How many years would it take for the lake to be totally covered?

Use with *Ups and Downs,* page 44.

11. Cut out the leaves. Paste them on another sheet of paper so that their centers overlap. How can you tell that the radius of the leaf does not grow linearly?

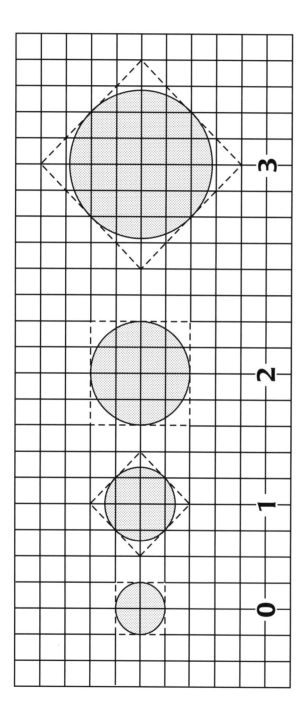

Week

MAKE UP A STORY

Use additional paper as needed.

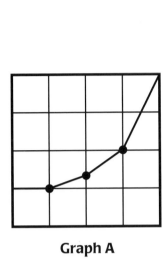 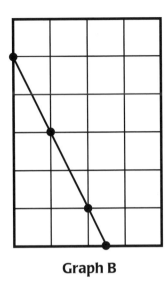 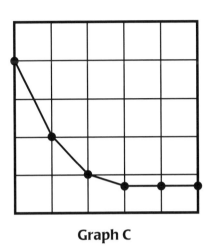

Graph A   **Graph B**   **Graph C**

1. These three graphs each tell a story about a situation in which something increased or decreased. Describe the pattern of increase or decrease for each graph.

2. Choose one graph and copy it on a sheet of paper. Make up a story that goes with that graph. Choose appropriate scales and labels for the horizontal and vertical axes, and include them on your graph.

DOTS, DISTANCES, AND SPEED

Use additional paper as needed.

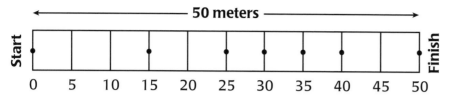

Here you see a strip that shows the movement of a car during a six-second time period. Every dot indicates the place of the car along the road. Between any two dots, the amount of time is one second.

1. What is the total distance the car covered in the first four seconds?

2. Draw a graph that shows how the distance increased over time.

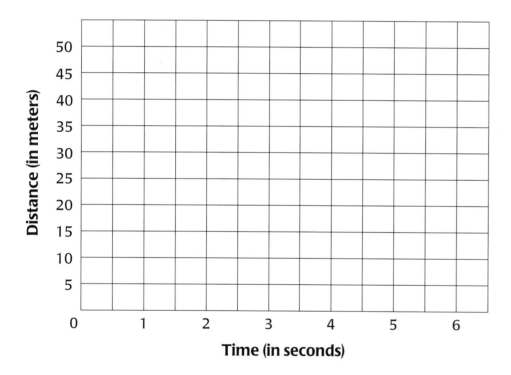

3. When did the car have the fastest speed? Explain.

4. Describe the pattern of increase or decrease in the above graph in your own words.

Use additional paper as needed.

This graph shows
how the water depth
in a coastal harbor
changes over time.

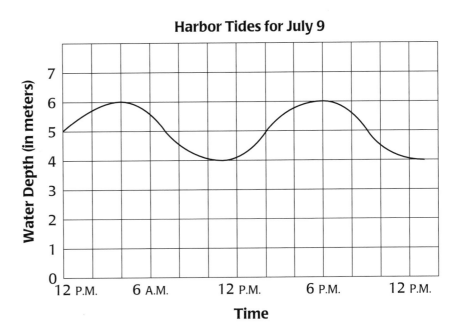

Harbor Tides for July 9

1. Why is this a periodic graph?

2. Color one cycle of the graph.

3. How long is one period of the graph?

4. A captain wants to know at what time on July 10 the water will be at its maximum depth. How can he find this out? What is the answer?

LIGHTHOUSE

Use additional paper as needed.

Ships without radar or other modern equipment on board can find their way at night by using light signals from lighthouses. Every lighthouse has its own light signal, so a captain can identify a lighthouse from the pattern of the light flashes.

Below you see a graphical representation of one lighthouse's signal pattern.

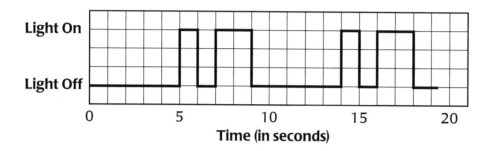

1. Examine the diagram of the lighthouse signal pattern. Shade the diagram to show one cycle of the signal pattern. How long is one period?

2. How many light flashes would you see in a one-minute period?

3. Create your own lighthouse signal pattern. Choose the number of light flashes and the period of the lighthouse signal pattern by yourself. Shade your diagram to show one cycle of your signal pattern.

TAXI!

Use additional paper as needed.

In the small town of Bakersfield, there are two taxicab companies. The Metro Taxi Company charges customers a flat rate of $4 per mile. The Acme Taxi Company uses the following table to show their cab fare costs.

Acme Taxi Company Fare Costs

Distance (in miles)	1	2	3	4
Cost (in dollars)	7	9	11	13

1. Draw a graph for *each* taxi company showing the costs of cab rides for various distances.

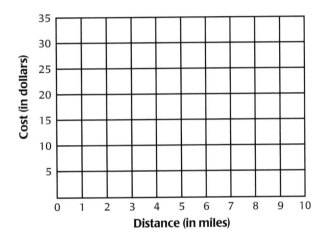

2. Write formulas that could be used to calculate the costs of cab rides using the two taxi companies. Use the letter *C* to represent the cost of the trip and the letter *M* to represent the distance in miles.

3. **a.** Write a proposal to the owner of the Acme Taxi Company, suggesting how she might increase profits by raising taxicab fare rates a little while still staying competitive with the Metro Taxi Company. Explain your plan in detail.

   b. Write a formula that could be used to calculate the cost of a cab ride using your new cab fare rates for the Acme Taxi Company.

GROWTH

Use additional paper as needed.

The table below can be used to show different types of growth patterns.

Time (in years)	0	1	2	3	4
Length (in mm)	10	20			

1. a. Complete the table to show a linear growth pattern.

   b. Write a NEXT-CURRENT formula that describes this situation.

   c. Write a direct formula that describes this situation. Use the letter *T* to represent time and the letter *L* to represent length.

2. a. Complete the table below to show a growth pattern that increases according to a certain growth factor.

Time (in years)	0	1	2	3	4
Length (in mm)	10	20			

   b. Write a NEXT-CURRENT formula that describes this situation.

3. Describe the shapes of the graphs for both situations: that with the linear pattern and that with the pattern that increases according to a certain growth factor.

BACTERIA IN FOOD

Use additional paper as needed.

The bacteria salmonella often causes food poisoning. At 35°C, a single bacterium divides every hour. Suppose there are 100 salmonella bacteria in a portion of food, and the temperature is 35°C.

1. Complete the table below to find out after how many hours there will be more than 1,000 bacteria present in the food.

Time (in hours)	0	1	2			
Number of Bacteria	100					

2. On a sheet of graph paper, make a graph that shows the growth pattern of the bacteria in the above table.

3. What is the growth factor for the situation described in problems **1** and **2?**

Suppose that by refrigerating the food, you are able to slow the growth of bacteria to exactly one-half of the growth rate in the previous situation.

4. Make a graph that shows this new growth pattern. Draw the graph on the same coordinate grid system that you used in problem **2.**

5. Estimate the growth factor for the new growth pattern described in problem **4.** Explain how you found your answer.

Use additional paper as needed.

Most nuclear energy is produced by splitting atoms of uranium. This atom-splitting releases great energy that can be used for many purposes. However, the atom may split in different ways to create as many as 200 different products! These products are called _nuclear waste._

Nuclear waste is highly radioactive, and it is therefore unsafe for living things. Nuclear waste must be kept in a safe place away from people, animals, and food and water supplies until it loses its dangerous level of radioactivity.

Suppose that a nuclear plant produced radioactive waste that would lose its radioactivity as follows: every ten years, the amount of radioactivity decreases by 25%.

One scientist suggested that the nuclear waste be safely stored in barrels for a period of 40 years until it is no longer radioactive. Then the barrels could be dumped into the ocean without causing any harm to humans, marine life, or food and water supplies.

1. How might the scientist have determined that it would take 40 years for the nuclear waste to lose its radioactivity? Do you agree with her plan? Explain your reasoning.

Use additional paper as needed.

Every 10 years, the amount of radioactivity in the substance decreases by 25%. This gradual decrease can be illustrated by drawing pictures. The picture below represents an initial amount of nuclear waste.

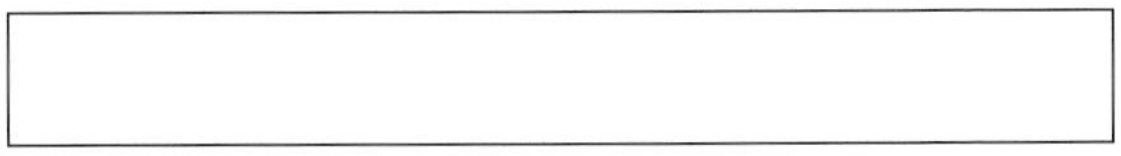

2. Using the above drawing as a model, draw a picture below to show how much of the initial amount of radioactivity is left after 10 years. Label this picture "Ten Years Later." Then draw another picture to show how much of the remaining amount of radioactivity is left after 10 more years. Label this picture "Twenty Years Later."

3. Carefully examine the two pictures you just drew that show the amount of radioactivity left after 10 and 20 years. Look back at your answer to problem **1.** Do you agree with the scientist's plan now? Explain your reasoning.

4. Now imagine a different situation in which a scientist started with nuclear waste material that showed a radioactivity level of 1,024 counts per minute on a Geiger counter. Make a table and a graph on a sheet of graph paper to show how the amount of radioactivity would change over time.

5. Give an estimate of the amount of radioactivity that still remains after 40 years.

6. Compare your answer to problem **5** with your answer to problem **1.** Do you want to revise your answer to problem **1?** Explain.

7. Write a NEXT-CURRENT formula to describe the decreasing amount of radioactivity in this situation.

Section A. Differences in Growth

1. a. The population grew by about 180,000 people. Estimates will vary. Accept estimates in the range of 170,000–190,000 people.

    b. Answers will vary, depending on students' estimates for part **a.** Based on a population growth between 1930 and 1940 of 180,000 people, the population in 1940 was 1,743,396 people (1,563,396 + 180,000 = 1,743,396).

2. a. The population grew the most between 1981 and 1990. Explanations will vary. Sample explanation:

    The graphs show that the greatest change in population, indicated by the longest bar, was during the decade ending in 1990.

    b. The population grew the least between 1931 and 1940. Explanations will vary. Sample explanation:

    The smallest increase in population was during the decade ending in 1940, and is indicated by the shortest bar on the graph.

3. a. The population grew the most between 1970 and 1980.

    b. Explanations will vary. Sample explanation:

    The graph line is steepest during this time period.

4. a. The population growth started to slow down after 1980 or between 1980 and 1990.

    b. Explanations will vary. Sample explanation:

    The graph shows a relatively flat line during this time period.

5. Answers will vary. Students' responses should indicate that Alabama's population increased at a steady rate for many years, then showed a big increase, and leveled off after 1980. Washington had a burst of growth in the 1940s that then slowed in the 1950s, but population growth has been steadily increasing since the 1960s.

6. Answers and explanations will vary. Sample response:

City 2 will have a larger population in the year 2000 since it has had the greatest increase in the last 10 years, 2.1 million to 2.5 million, and that increase was four times as much as in the decade before.

Section B. Cycles

1. Answers will vary. Sample response: Before the oven is turned on, it stays at room temperature. After the oven is turned on, the temperature rises rapidly until the oven is shut off. Then the temperature begins to decrease as the oven cools down.

2.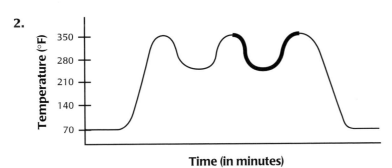

3. See the graph in the solution to problem **2.**

4. See the graph in the solution to problem **2.**

5. **a.** Answers will vary. Some students may label the horizontal axis in 5-minute increments. So, one cycle would be 10 minutes long.

   b. Answers will vary, depending on students' answers to problem **3.** For one example, see the graph in the solution to problem **2.**

   c. Answers will vary, depending on students' answers to problem **3.**

Section C. Linear Patterns

1. 110 centimeters

2. a. Add 15 centimeters to the current height of the water.

   b. NEXT = CURRENT + 15

3. It will take 6 hours and 40 minutes to fill the pool to a height of 180 centimeters. Explanations will vary. Sample explanation:

   Mark wants to raise the water level of the pool to 180 centimeters. The current water level is 80 centimeters, so he needs to raise the water level 100 centimeters. The water level rises 15 centimeters per hour. I divided to find the answer.
   $100 \div 15 = 6\frac{2}{3} = 6$ hours 40 minutes.

4. See the graph below. Since Mark has been filling the pool for three hours at 15 centimeters per hour ($3 \times 15 = 45$ cm), the graph for filling the pool with two hoses starts at 80 cm + 45 cm, or 125 centimeters.

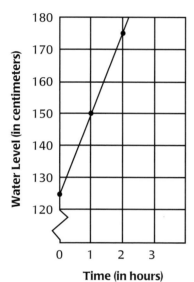

5. a. *H* stands for the height of the water. *T* stands for the amount of time in hours that two hoses have been filling the pool.

   b. $H = 125 + 25T$

   c. $H = 125 + 15T$; *T* would be multiplied by 15 instead of 25.

   d. Mark saves a total of 1 hour and 28 minutes. (Using only the first hose, it would have taken 6 hours and 40 minutes. Using two hoses after the first three hours, it took 5 hours and 12 minutes.)

Section D. Faster and Faster

1. The amount of bacteria will be over the limit in a little less than 9 hours.

Hours	0	1	2	3	4	5	6	7	8	9
Number of Bacteria	200	400	800	1,600	3,200	6,400	12,800	25,600	51,200	102,400

2. The inspector would have found 20,000,000 bacteria one hour earlier.

The inspector would have found 80,000,000 bacteria one hour later.

3. NEXT = CURRENT × 2

4. The mousse was removed from the refrigerator almost 9 hours before it was inspected. Strategies will vary. Some students may count how many times they must divide by 2 until the quantity of bacteria is under 100,000.
(40,000,000 ÷ 2 = 20,000,000 ÷ 2 = 10,000,000 ÷ 2 = 5,000,000 ÷ 2 = 2,500,000 ÷ 2 = 1,250,000 ÷ 2 = 625,000 ÷ 2 = 312,500 ÷ 2 = 156,250 ÷ 2 = 78,125).

Section E. Half-Lives

1. The temperature of boiling water after cooling down for 1 minute is 90°C; after 2 minutes 81°C; after 3 minutes 72.9°C (or about 73°C); after 4 minutes 65.61°C (or about 66°C).

2. a. This formula is not correct. Explanations will vary. Sample explanation:

The formula does not hold true when the CURRENT is 90; 90 − 10 = 80 and not 81.

b. This formula is correct. Explanations will vary. Sample explanation:

0.9 × 100 = 90
0.9 × 90 = 81
0.9 × 81 = 72.9

c. This formula is correct. Explanations will vary. Sample explanation:

100 − 0.1 × 100 = 100 − 10 = 90
90 − 0.1 × 90 = 90 − 9 = 81

d. This formula is not correct. Explanations will vary. Sample explanation:

0.1 × 100 = 10 and not 90.

3. It takes boiling water somewhere between 8 and 9 minutes to cool down to a temperature of 40°C. Explanations will vary. Sample explanation:

I used the formula NEXT = CURRENT × 0.9 and made a table to show my answers.

Minute	0	1	2	3	4	5	6	7	8	9
Temperature (°C)	100	90	81	72.9	65.6	59	53.1	47.8	43	38.7

Cover

Design by Ralph Paquet/Encyclopædia Britannica Educational Corporation.
Collage by Koorosh Jamalpur/KJ Graphics.

Title Page

Phil Geib/Encyclopædia Britannica Educational Corporation.

Illustrations

6, 8 Phil Geib/Encyclopædia Britannica Educational Corporation; **10** Brent Cardillo/Encyclopædia Britannica Educational Corporation; **12, 14, 18** Phil Geib/Encyclopædia Britannica Educational Corporation; **22** Paul Tucker/Encyclopædia Britannica Educational Corporation; **24, 34, 36, 38, 40, 44** Phil Geib/Encyclopædia Britannica Educational Corporation; **48** Jerome Gordon; **54, 56, 60, 66, 70** Phil Geib/Encyclopædia Britannica Educational Corporation; **74, 76, 78, 80** Paul Tucker/Encyclopædia Britannica Educational Corporation; **82, 84, 96** Phil Geib/Encyclopædia Britannica Educational Corporation; **106** Brent Cardillo/Encyclopædia Britannica Educational Corporation; **112, 114** Phil Geib/Encyclopædia Britannica Educational Corporation; **116** Brent Cardillo/Encyclopædia Britannica Educational Corporation; **136, 160** Phil Geib/Encyclopædia Britannica Educational Corporation.

Photographs

26, 28 © AP/Wide World Photos; **32** © David J. Sams/Tony Stone Images; **52** © David W. Hamilton/The Image Bank; **58** © Richard Heinzen/SuperStock, Inc.; **66** © DUOMO/Steven E. Sutton; **92** © Kenneth Mantai/Visuals Unlimited; **96** © George Musil/Visuals Unlimited; **98** © Dick Keen/Visuals Unlimited; **116** © John Gerlach/Visuals Unlimited.